Spinning for Softness & Speed

PAULA SIMMONS

THE BOOK MAN

Chilliwack, BC

Copyright 1982 by Paula Simmons
All rights reserved
Printed by Sunrise Printing Inc.
Chilliwack, BC, Canada

Published by
The Book Man
www.bookman.ca
604.792.4595

Fourth Printing 2009

Library of Congress Cataloging in Publication Data
Simmons, Paula
    Spinning for Softness & Speed
    1. Handspinning.   I. Title
TT847.S573   1982   746.1'2   82-8921
ISBN 0-914842-87-0

Contents

Spinning for Softness & Speed

1 What the One-Handed Method Will Accomplish

M uch of the pleasure of yarn spun by hand lies in its warmth and softness. The harsh and unpleasant effect of overtwist can defeat the whole purpose of handspinning. The one-handed method presented here will virtually guarantee a soft yarn.

In addition to conquering overtwist, the one-handed method brings with it several other benefits. It will increase your spinning speed dramatically. It will allow you to spin without looking at the yarn, and with less fatigue. The yarn you produce will be more evenly spun. And, finally, you will have learned a technique that, combined with others already in your repertoire, can handle wool of any type or condition.

CONTROL OF OVERTWIST

With this soft-yarn method, you can conquer overtwist in a day by training your hand to deal with it automatically. It is sometimes easier to train your hands than it is your head, for just knowing how to do something does not mean that you can do it.

With all other spinning methods, you can allow almost any amount of twist between your two hands as you spin—even differing amounts of twist each time you draw out a length of yarn. With fine yarn, the result

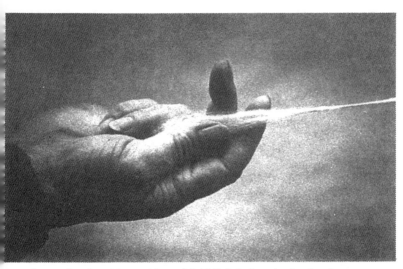

Correct hand position, with wool held lightly in palm by ring finger and little finger.

may be localized areas so overtwisted that the strain actually causes the yarn to break. There is nothing about two-handed techniques that in any way trains your hands to solve these three basic problems:

1. How much twist is needed.
2. How to limit the amount of twist.
3. How to keep that amount constant.

However, with the one-handed soft-yarn method, the amount of twist that you allow into the yarn is no longer left to your own discretion. Your hand draws out well ahead of the twist, and must keep ahead of it in order even to continue spinning. Any more than a medium amount of twist will stop you right in your tracks, as will be explained in the next chapter. The method not only produces a soft twist, but makes it possible to completely control the amount of twist in the yarn *throughout* the skein, keeping it consistent.

In spinning with one hand, you have, for any given fiber or yarn size, only a very narrow range of twist that will work. The minimum is enough twist to hold the yarn together; the maximum must be less than the amount that would prevent easy slippage of fibers in front of your hand, so that you would have to forcibly draw out the yarn (with both hands) rather than let the twist form the yarn for you. Thus it is easier to spin yarn

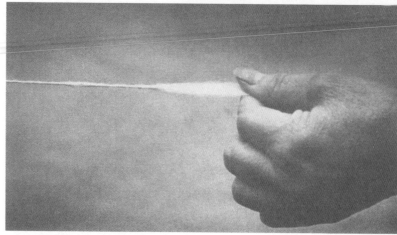

Wool held with thumb and first two fingers. This way you have to look when you spin.

with a constant amount of twist than it would be to spin yarn in which the amount of twist varies. This advantage alone could shortcut a year of practice in trying to attain a consistency of yarn size and twist from one skein to the next.

INCREASED SPEED

Unnecessary hand movements can stand in the way of speed and relaxed spinning. The one-handed method will virtually eliminate these.

Everyone develops certain mannerisms that can become habitual and unconscious. In classes these can be passed from teacher to students — if not by word, at least by example. Students in turn may develop further habits that do not really facilitate the process, and too often these movements cannot be speeded up.

You may have seen spinners who, especially when they are doing the short draw, twist the yarn with the hand nearer the orifice, as if to add twist. If this is brought to their attention, they will agree that it is quite unnecessary — the spinning wheel will do the same amount of twisting with or without the help of the hand. But once a habit is formed it is difficult to stop.

The discipline of spinning with one hand makes inefficiency impossible. The only habits you acquire are those which enable you to spin faster. When the whole operation is performed by one hand, it almost has to be done correctly. (I say "almost" because you *can* spin with one hand without holding the wool in the way suggested here. The only drawback of the different hand position is that it does not completely enable you to spin without looking.)

In addition, softly spun yarn spins up more quickly than firmer yarn. The fewer the twists per inch, the faster you can turn out yarn with the same amount of treadling.

SPINNING WITHOUT LOOKING

Once you get the feel of it, you will be able to spin without looking at the yarn or your hands. This is not a matter of showing off; it has practical applications. If you can spin without looking, you can then read a book, or watch television, or look out the window at the sheep. If you are holding a conversation, you can look at the person with whom you are talking. I have heard more than one knitter complain that her husband did not want her to knit if they were sitting around talking together. He had the impression that he was not being heard, or was not getting her undivided attention. The same can hold true if you are always looking at your spinning.

If your eyes are tired, you can rest them while spinning. Unfortunately, though, if you are sleepy you can go to sleep while spinning. Seriously, not having to watch the yarn can give you a more relaxed feeling, without having to slow down.

It was over forty years ago that I first became aware that it was possible to spin without looking. I learned of the Hand-spinners for the Blind, who were in New York at that time. It was claimed (and this really intimidated me then) that blind spinners could each take two yards of mill roving and spin exactly the same number of yards of yarn from it, with only the experience of their hands to guide them. The skill was in their hands, not in their eyes.

EFFORTLESS SPINNING

At first it is gratifying to spin almost any kind of yarn, however awkwardly it is done. But since speed and efficiency are often tantamount to style and grace, the sooner bad habits are corrected the more productive and enjoyable and effortless spinning becomes.

This soft-yarn method allows a relaxed freedom of movement

which is not tiring, even over a long period of time. In spinning for hours you might get bored, but not physically tired.

You can also enjoy the spinning process more when you are not fighting the twist or wearing yourself out trying to keep a constant size of yarn from skein to skein. Spinning should be, and can be, a very relaxing activity — but only if it can be done effortlessly and without frustration. While you may feel frustrated the first time you try to spin with one hand, a little practice can eliminate some other frustrations that spinners customarily encounter: overtwist, difficulties in maintaining constant size and twist from one skein to another, and problems with "difficult" wool.

This method trains your hands to operate as automatically as your feet do, regulating size and twist almost by instinct. With this developed sense of touch, spinning becomes less of a manipulation and more like a caress.

SMOOTH EVENLY SPUN YARN

An understandable goal is to want to attain the skill to produce evenly spun yarn at a decent rate of speed. With the one-handed method, you can easily reach that goal, provided you have a spinning wheel that works well.

Fiber requirements for smooth, even yarn are:

1. Clean wool, well washed, or a fleece that is so clean it does not need washing (see Appendix for washing information).
2. Absence of seeds and vegetable matter in the wool, as these can cause slubs.
3. Wool well teased or picked, and carded, preferably with a drum carder.

Fine yarn is the easiest size to learn to spin evenly, and this technique is the fastest way to spin fine yarn. The faster you spin, the smoother it will be.

VERSATILITY

The one-handed method can be combined with other techniques, especially the unsupported long draw (see Appendix), to give a high degree of versatility. The ability to handle fiber of any type or condition can be very useful. This is particularly true if you have to teach others, with their wool and on their wheels.

Any *one* technique, no matter what it is, will not really equip a spinner to cope with all the problems that can be encountered. Short wool, long wool, dry wool or especially sticky wool, well-carded or insufficiently prepared wool—all present a different challenge in spinning. The soft-yarn method can usually be the *basis* for handling the fibers; combined with any of several other techniques (see Chapter 5), it will give complete yarn control whatever the problem.

Spinning for
Softness &
Speed

2 How to Spin
with One Hand

Proper fiber preparation is essential in learning the one-handed spinning method. Grease wool is credited with holding together better, which sounds like an advantage for beginners. With the soft-yarn method, however, this is actually a disadvantage. Here, you can control fiber slippage more reliably with washed wool, because the stickiness or gumminess of grease wool can prevent the smooth, free slippage of fibers that is necessary when you spin with one hand. Any condition that impedes this smooth slippage is a disadvantage with this method, and not only makes it more difficult to spin with one hand but slows you down. While it is possible to occasionally find a fleece that spins this well without washing, such fleeces are unusual. (See Appendix for washing instructions.)

After your fleece is washed and dry, take a lock of the wool and just draw it out between your hands. If you have washed it well enough to remove the gumminess but the fibers still do not slip against each other smoothly and easily, try a light misting with warm water containing about 1/3 liquid laundry anti-stat. Wrap up a few hours in a warm place before carding and spinning.

The one-handed spinning method *can* be done with

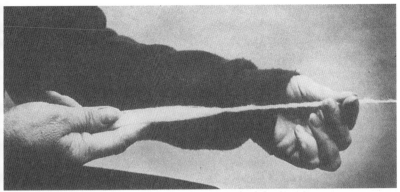

If you spin this way, you will use your right *hand to hold the wool.*

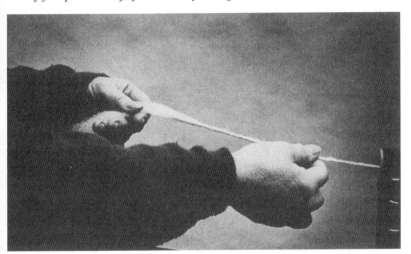

If you spin this way, you will use your left *hand to hold the wool.*

Correct position: wool rests in palm.

almost any kind of wool, but when you are first learning it, you will want to make it as easy as possible for yourself. Try out a few different wools, if you have a choice, to see which spins with the least difficulty. Don't be tempted to practice with difficult wool just because it is so expendable. You can get discouraged too quickly. Later, when you are a more versatile spinner, this same wool may not be such a problem.

The same type of wool will not be equally easy for everyone, but in general wool of a medium length, not too fine, washed clean of all gumminess and well carded, will work best. (Carding is discussed in the following chapter.)

SPINNING-WHEEL ADJUSTMENT

It may be necessary to adjust your spinning wheel differently from the way you would have it for other methods of spinning.

With a single-belt wheel you may need to adjust the brake band. Adjust it a little at a time until it pulls sufficiently, but not so much as to pull a soft yarn apart. Beginners often make the mistake of adjusting the band too much at a time, which then results in much back and forth adjusting—too much, too little, too much—before finding the right tension.

If you are using a double-belt wheel, try a pulley size (if you have it) with up to a 2:1 ratio, but with a fairly loose drive belt. Experiment with the drive belt tension until it is just firm enough to draw against, but not firm enough to be a strain on soft yarn. If you do not know the pulley ratio, just use the largest flyer pulley with the smallest bobbin pulley.

Keep in mind that softly-spun yarn still gets a little more twist added to it as it is pulled onto the bobbin, and will be stronger than you expect.

HAND POSITION

The soft-yarn technique is done by spinning with one hand. In deciding which hand to use, you need only note which hand you now hold further from the orifice when you spin: that is the hand that will hold the fiber supply for this method. The hand that is ordinarily nearer the orifice will here be called your "free" hand or your "other" hand, and you should try not to use it in this spinning process. Your eventual goal will be not to use it *unnecessarily*.

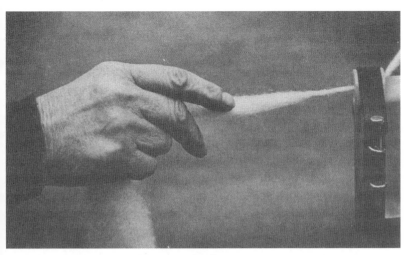

Faulty position: wool hangs out of palm.

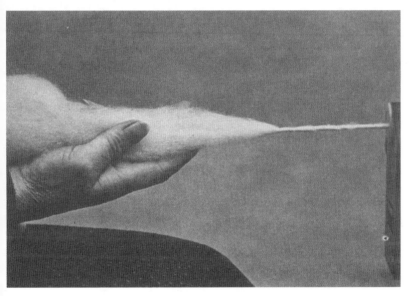

This is too bulky a wool supply for the beginner.

In order to spin with only one hand, your fiber supply must be held *very* lightly. You want the twisting action of the spinning wheel to be able to draw out fibers from those held in your hand and form them into a soft yarn as you draw your hand back, drawing back always ahead of the twist.

The hand holding the fibers should be positioned with the palm *up*, at least while learning, so that the wool rests in your

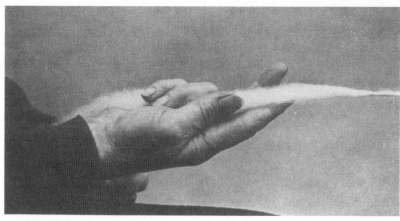

A thinner mass makes it easier to feel what is happening.

palm. This is where you will want to be able to feel the slippage of fibers as you spin. It is your palm that will be most aware of fiber movement, and this method of spinning trains it to be even more sensitive in a relatively short time.

FIBER SUPPLY

The fiber mass, when you are beginning to learn, should be about the size of a hand-carded rolag. Too bulky a fiber mass, at the start, prevents you from feeling the slippage of the fibers clearly. Once you have become used to this feeling, the size of the fiber supply will not matter as much.

Some wools actually seem to spin better from a dense mass, while others require a more elongated, thin mass. What it amounts to is that the easier the wool is to spin, the less the density of the fiber supply held in your hands matters. Any time you start to spin a new fleece, experiment with a slender mass and a thick mass to see which works best.

While you may not want too bulky a fiber mass when you are learning, it is not necessary to have an *even density* of fibers. They will be drawn out just as evenly when the fiber supply is not uniformly dense, as long as the wool is clean and not gummy.

YARN APPEARANCE

The main thing while learning is *not* to be concerned about how your yarn looks, for the present. Your purpose now is not to produce yarn but to train your hands, to acquire certain habits. Once the right habits are formed, the result will be a yarn that is pleasantly soft and of the very size you intend it to be.

While you should not be preoccupied with the appearance of the yarn while learning this method, mastering the method does assure complete control over size, twist, and texture. Spinning with one hand, where you must stay ahead of the twist with the particular wool you are using, makes it possible to maintain an almost uniform amount of twist throughout each skein because of the narrow range of twist possible with this method of spinning. This uniformity of twist, in turn, helps regulate the yarn size. I would call it a medium or average amount of twist, because it lies in the range between the minimum twist needed to hold the yarn together nicely and the maximum twist possible without halting the fiber slippage.

SPINNING

With your hand turned upward, hold the wool against your palm with your ring finger and your little finger, which have a naturally light touch. Try *not* to touch it with your thumb and first two fingers. Your thumb and fingertips are not the most sensitive parts of your hand; they are not where you get blisters if you are digging in the garden. It is in your palm that you get the blisters, and that is where you have the most feeling.

You can attain the same kind of soft yarn by holding the fiber supply with your thumb and first two fingers, but you lose some of the advantages of the soft-yarn method. When you use these fingers, you are not free to spin completely without looking, because they lack the sensitivity of your palm. Also, if the yarn starts to pull apart from undertwist, you cannot catch it as easily. With the fibers in your palm, those fingers are out in front, ready to clamp down on the yarn if it threatens to get away. This sounds difficult, but with practice it becomes almost a reflex.

The hand holding the fiber supply should draw back just ahead of the twist that is being inserted as you treadle. It will keep a slight tension on the yarn that is being drawn out. This tension is necessary to keep the twist going into the fiber supply and to keep the yarn from winding onto the bobbin until you want it to wind on. Actually, if you are drawing out at the correct speed, this will automatically be keeping an even tension on the yarn being spun, until you allow it to wind onto the bobbin.

While the fibers are being drawn out of the fiber supply by the twisting and pulling action of the spinning wheel, your hand must retreat away from the twist, keeping well ahead of it at all times, as you continue to treadle. When you have drawn your hand back in a comfortably long sweep, let the spun yarn be pulled onto the bobbin. If you hesitate at the end of the draw before allowing the yarn to pull onto the bobbin, you will be adding more twist. There may, of course, be instances where you will want to add more twist, depending on the intended use of the yarn, but for soft yarn the twist should be kept to a minimum.

YARN TWIST AND SIZE

With some techniques the yarn is still being attenuated after it has some twist in it. However, in the one-handed method, the yarn is completely spun and its size determined as it emerges from the fiber supply, drawn out only by the twisting action from the spinning wheel. The secret of maintaining the one-handed action is to stay ahead of this twist at all times. If your spinning wheel is functioning well, it is not possible to over-twist unless you deliberately hesitate at the end of each draw to insert more twist.

As your hand becomes more sensitive to the feel of a soft twist, it will at the same time become attuned to the coordination between hand and feet that accompanies this kind of twist. This will become automatic in a surprisingly short time.

Do not be disconcerted if the yarn gets progressively finer when you are practicing. With this system it is easier to do fine yarn than heavy yarn. Also, it is probably the fastest way to spin an evenly spun fine yarn. With most wools, a medium-heavy yarn spun with this one-handed method will be more irregular than a finer yarn spun from the same wool. But with any given wool you will find that there is a certain degree of fineness or thickness that is most easily spun. When you are just practicing, the size which spins most easily is the size to do.

Yet, while you are learning, try not to be critical of either size or texture. Once you can manage the hand action, the control will follow.

LIGHT TOUCH

If you come to the end of the fiber supply in your hand and have a snarl that you cannot spin, this is because you have been holding on too tight. This has an effect on the wool like back-combing your hair, leaving it tangled at the end. Keep a light touch.

USE OF THE THUMB

Remember to keep your thumb off the wool. Try pointing your thumb and first two fingers toward the approaching twist, but hold the wool (lightly, remember) in your palm with your ring finger and little finger. This keeps the thumb off the wool and also gives you a way of catching the yarn with those extra fingers if it starts to get away from you due to undertwist.

In actual yarn production (as opposed to practicing) you will be using your thumb quite a bit, as a control factor in spinning, but for now it is important that you train yourself *not* to use it. As a result of this training, you will use it only as needed.

With wools that do not draw out smoothly, you will find that you can still hold them with the little finger and ring finger against your palm, but may need to do a pinching action with your other fingers. This holding and releasing, as needed, can counteract the intermittent jerkiness of the wool slippage and keep it from affecting the yarn size and twist.

CONTROL OF THE TWIST

The first misgiving to be overcome in using this method of spinning is that the wool will separate due to lack of twist — that it will not actually hold together as yarn. This is exactly what does happen at first. However, in just a matter of hours your hand will begin to sense any thinning out of the fibers. This thinning out indicates either that the yarn is becoming thinner or that the yarn is pulling apart because the hand is too far ahead of the twist. This awareness is the signal for your hands to draw out slower. If you do not sense it in time, you will need to use the thumb and index finger to clamp down on the fiber supply momentarily while more twist builds up. This will strengthen the already spun yarn and make more twist available for the ready-to-be-spun fibers that have been compressed by thumb and index finger. As you let go with your thumb, you will find that you can now start to draw out the yarn as before.

If this tendency toward thinning and pulling apart is felt at the end of the long sweep of drawing out, by the time you allow the yarn to be drawn onto the bobbin the problem may be solved by the extra twist that is added as it is drawn in. In this case, you will allow it to draw on a bit slower than usual, to add a little extra twist. And, as you move your hand to get ahead of the twist again for the next length of drawing out, you can now go at a speed commensurate with the amount of twist for that size of yarn. The feel of what is happening, even before you see what is happening, is what must be your guide. All that is needed is more practice in recognizing the feel in order to make the appropriate response.

If your hand does not stay ahead of the twist, the twist passes your fingertips and enters into the palm-held fibers, which blocks the spinning process. If you keep trying to draw out, the fibers will not slip and you will just be pulling the yarn off the bobbin. You could use the other hand to pull against, but the point of using one hand is to force you to stay constantly ahead of the twist, which in turn restricts the amount of twist. What you should do is slow up your treadling and move your hand back quickly, grasping the fiber supply further back. This lets you get ahead of the twist once more. While you are learning, you may have to revert for a minute to the use of both hands until you get the yarn under control.

In trying to learn this one-handed technique, a spinner will sometimes allow the yarn to be drawn out from a large fiber mass that is in *front* of her hand. While this is not a bad or wrong technique, it is not quite what is meant here. I should mention that it is more suitable to heavy yarn, for when used with fine yarn it makes it too easy for the yarn to have less twist than needed, and fall apart.

The point of contact between the twist from the action of the spinning wheel and the unspun fibers that are being drawn into the twist will be just beyond the tip of your index finger, as it is pointed toward the orifice. With short wool, the twist will come a little closer; with long wool, it will be further away. This is sometimes expressed as "a short drafting triangle" or "a long drafting triangle."

This may be a good time to mention the use of the terms

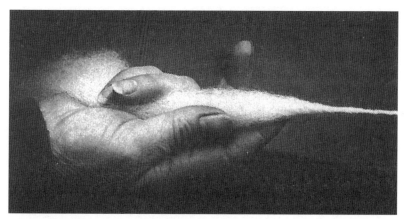

Medium-short wool makes a short triangle.

"drawing out" and "drafting," which are often used almost interchangeably. It seems more accurate to speak of "drawing out" in reference to the hand's retreating ahead of the twist, with the fibers being drawn out of the fiber supply by the twist, and of "drafting" as the attenuation of fibers during the twisting, which requires both hands.

EFFORT

Spinning with one hand requires a minimum expenditure of energy, and its effortlessness will be apparent the first moment you get the feel of it. I've had spinners say, with amazement, "I'm not doing anything, the spinning wheel is doing all the work!" And so it is. Once the palm of the hand is trained to control the twist by the feel of the fibers as they move, and draws back consistently ahead of the twist, spinning does become virtually effortless. This can be important if you plan on selling yarn or spinning large quantities for weaving, both of which entail long sessions of spinning. Try to maintain a relaxed body position and let the wheel do your work for you.

SPEED

Most spinners can spin faster than they think they can, or faster than they have tried to spin. Speed itself puts more pressure on your hands to solve the problem. With more pressure, you are more prone to think with your hands rather than patiently trying to translate thoughts into an appropriate act. Try to spin a little faster than you think you can, and you may be pleasantly surprised. Another reason to spin a little faster is that at a slow

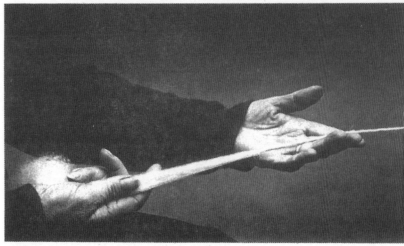

Longer wool makes a longer triangle.

speed there is so little action of the fibers slipping that it is rather difficult to feel anything happening, while there is a very definite feeling if they are slipping more quickly.

Impediments to spinning speed can be a slow or faulty spinning wheel, poorly prepared or sticky wool, inefficient hand movements, or preoccupation with a perfectly smooth texture. (More about speed in Chapter 4.) Hand movements will become more fluid with this technique, and if you are working with well-prepared wool you will be able to do an evenly spun yarn at high speed.

THE OTHER HAND

You may need your other hand to help out from time to time, but while you are practicing, resist using it to improve the yarn size or texture. Once you get past the practice stage with this method, you *will* be using the other hand occasionally, and also combining this technique with others (see Chapter 5). Obvious uses of the other hand are reaching for additional wool and picking out seeds or bits of vegetation.

With long wools, you may find that you need to use the other hand to pull against (as in the long draw), but only at the very end of each draw. The final brief pulling can be crucial to determine the yarn's texture because of the long fibers. With long wool you will find that it is necessary to keep the hand holding the fibers further ahead of the twist than with shorter wools. This is especially so in spinning fine yarn with long wool, for you will be working with a belt tension that is not quite firm enough to pull against. Long wool is especially tricky to spin with a soft feel, for the amount of twist necessary is so much less than is needed when spinning short fibers.

Some spinners hold their free hand open and flat under the yarn, as seen in the photo of the long drafting triangle (page 19). They have more confidence this way, feeling that the free hand can be used to grab the yarn if it gets away. It can also serve another purpose. With some wheels, such as bulk spinners, this hand under the yarn acts as a kind of brake and helps counteract the jerkiness of the yarn wobbling in the large orifice.

In actual production of yarn, you won't really be spinning a whole skein without ever using your other hand. (More about that in the next chapter.) All I want to say here is that it is no great disaster if you do have to use both hands occasionally. Just aim for familiarity with how it feels to spin one-handed. Cultivate the feeling and you will soon learn the technique.

*Spinning for
Softness &
Speed*

3 Wool Preparation and Handling

The one-handed method of spinning soft yarn can be learned in a day, given certain conditions: a good working wheel and fibers that are clean and well prepared.

CLEAN WOOL

For learning this particular technique, clean wool is easier to work with than greasy. Grease wool does not draw as evenly nor as smoothly. When washing wool, I use primarily a soaking method, soaking fleeces in quite hot water with *lots* of detergent. (See Appendix for detailed instructions on how to wash wool.) The washing is intended mainly to remove the gumminess that interferes with the smooth slippage of fibers when spinning. If a fleece feels too dry after washing, or been washed and stored, some spinners use water-soluble spinning oil.

An alternate way of "conditioning" is a light misting with warm water containing liquid laundry anti-stat such as Downy®. Wrap wool up over night, for carding the next day.

Occasionally someone asks about yarn spun "in the grease" so that it will be water-repellent. It should be pointed out that in order to obtain preshrunk yarn for a garment that can be washed later on, you will need to pre-wash wool and yarn. In any case, even after it is

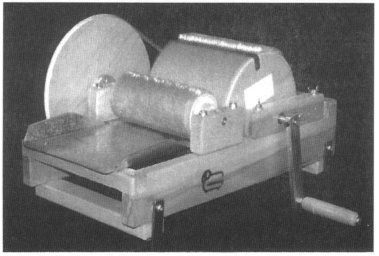

*Deb's Delicate Deluxe, a versatile drumcarder from
Pat Green Carders Ltd.*

washed, clean *undamaged* wool has a thin water-repellent skin,
the epicuticle, which gives it considerable resistance to wetting.
(This is not necessarily true of mill-produced yarn, for if yarn is
treated for shrink resistance by chlorination, the epicuticle is
removed and the yarn becomes more wettable.)

CARDED WOOL

It is easier to learn to spin with wool carded on a drum carder
than with hand-carded rolags. Yet, after you have learned, both
will be equally easy to spin. I do discourage the use of hand
cards any more than is absolutely necessary. Several people I
know have required surgery on the tendon sheaths of their
thumbs from the overuse of hand cards. Hand cards are most
useful in demonstrating at fairs or colonial restorations, and in
showing basic wool skills to children.

A good drum carder, handled properly, should last a lifetime,
lessen the risk of thumb injury, and prepare your fibers nicely
for easy spinning. To minimize wear and avoid damaging your
carder, always have your fiber well teased, either by hand or
with a picker. (See Appendix for picker use.)

With batts from a drum carder, I find it faster to tear the batt
into a zig-zag strip for spinning, rather than tearing it into
separate strips or attenuating it into one long thin strip. The
difference between the first two options is slight. I admit a
prejudice against the attenuating of the whole batt into a long

narrow strip because this more closely approximates the factory method of condensing fibers into a thin roving during the process before the final spinning. In factory methods, the final drafting will be to only about twice the length of that roving (half its size). With handspinning, the hand-held mass can be from five to fifty times the diameter of the finished yarn, and much of the distinctiveness of handspun yarn can be the result of this ratio.

SHORT WOOL

In spite of the general feeling that a long staple is more desirable for spinning, do not rule out the virtues of short-fibered wool. Short wool, if clean and well carded, is very good for one-handed spinning, especially into fine yarn.

Short lambswool has quite a different feel from the fleece of a mature sheep of a short-wool breed, and wool of each separate breed will have its own distinct qualities. The feel of the yarn will vary, depending on the softness or crispness of the wool. Only for spinning thick, heavy yarn is short wool less suitable.

LONG WOOL

When first learning, you'll find that long wool is difficult to spin, particularly into soft yarn. With two-handed spinning techniques, the hands must work quite far apart. With this one-handed method, the hand must stay further ahead of the twist, as less twist is needed to hold the yarn together. For better control of yarn size when spinning long wool, it is often necessary to assist with your other hand at the end of each drafting. This gives a firmness against which you can pull to even out the yarn size or draw out any unwanted slubs.

Opinions about the uses of long wool vary. In some countries it is considered to be suitable only for carpets; in other countries it is not carded but only combed, and then made into worsted yarn.

MEDIUM-LENGTH WOOL

There is less disagreement about the properties and uses of medium-length wool than about short wool or long wool. Even spinners who are devoted to one extreme or the other will describe their preference as short-to-medium, or medium-to-

Light use of thumb and index finger in spinning wool that has been picked but not carded.

long wool. To that extent, medium-length wool could be considered ideal.

PICKED WOOL

Wool does not *have* to be carded for spinning into soft yarn. There can be a disadvantage in wool too meticulously prepared. Yarn spun at top speed from nicely prepared wool can be spun so evenly that it looks more like mill yarn than like handspun. This can be considered an advantage if you want smooth, perfect yarn, or a disadvantage if you want something with more irregularity. Sometimes proficiency itself can be a handicap.

Thus the teasing action of a hand-operated wool picker (see Appendix) may be adequate preparation for a nice fleece. This preparation has the advantage of allowing a more varied texture and color than you are apt to get if you are using well-carded wool. Like something handwritten as opposed to something typewritten, yarn from wool not overly worked has a more "autographic" appeal than that which has been too mechanically processed. Like other craft products, interesting yarn is frequently identifiable. Having seen the yarn, we recognize the spinner.

With uncarded wool — picked or teased — the technique for spinning is slightly different. Though you will still hold the fibers lightly in the palm of your hand, you will make use of your thumb and index finger, which ordinarily are best kept off the wool. The wool is still being drawn out from your palm, but it is

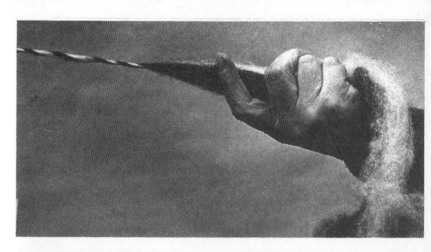

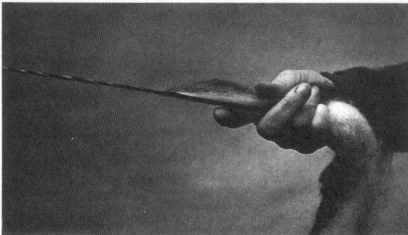

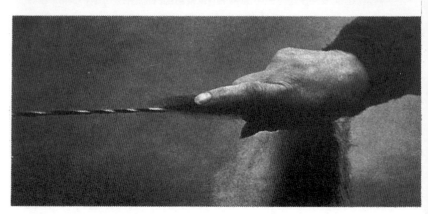

Hand turning back and forth to facilitate the spinning of two different colors into variegated yarn.

25   Wool Preparation and Handling

The unsupported long draw, stretching the fibers between two hands, forces the drafting together of two fiber types.

lightly compressed as it passes between thumb and finger, thus giving you more control over the amount that enters the twist. This light pressure also stretches out some of the uncarded fibers into somewhat the same arrangement they would have if carded, allowing them to be more easily drawn into the twist.

WOOL LAYERED IN CARDING

If you find it unpleasant to work with underprepared wools in order to get more texture, you can layer two different shades or two different wool types on the carder and spin both at once for a more interesting appearance.

The layering of two different colors in carding does not necessarily affect the texture, only the appearance. However, if these two different shades are also differing wool types, then it does cause a certain problem that does affect both the texture and appearance.

The tendency, as you spin this layer of different wool types, is for the longer wool to be selected and caught up more readily and drawn into the twist, leaving you with the shorter of the two wools still in your hand. To counteract this tendency, turn your hand back and forth to manipulate the shorter fibers into a position where they too will be caught up into the yarn. For this you may have to watch while spinning.

A layered batt of two different wool types can usually be spun with a more spiraling effect of the two colors by using the unsupported long draw. This consists of releasing a visually measured length of wool fiber mass and drawing that out into yarn before releasing more of the two-layered colors. Which of the two techniques you use may be determined more by the yarn size than by the fibers. With a fine yarn, you would still do best with the soft-yarn method, possibly combined with the unsupported long draw, and using a thinner mass of fiber supply. For medium-sized yarn, the unsupported long draw should be more efficient and give a better spiraling of colors.

CUSTOM-CARDED WOOL

Custom carding by a woolen mill has the disadvantage of damaging wool fibers. This fiber breakage results in shorter wool than you had before carding. Yet this mill-carded shorter wool draws out evenly and spins up well using one hand, so it is good for learning this method. It can be spun into nice warp yarn, and the evenness of the machine carding makes it possible to spin it very fast. See the advertisements in *SpinOff* magazine to locate custom carders who specialize in carding fleeces for handspinners.

LAYERING AND CUSTOM CARDING

A more real disadvantage of custom carding is that no matter what mish-mash of odd shades and wool types you send, you frequently get back one medium-color batt of a nondescript gray.

But there is a way of getting around this. Save up your odds and ends of problem wools for custom carding, keeping them in three separate boxes. (This is easier than bags when saving up; just toss the wool at the box.) No need for a really careful sorting — just light, medium, and dark. When you bag the wool to send it out (and this is really important), mark each bag very plainly, *"Card This Bag Separately."* Mill workers are being paid to card, not to think about it. If your name is on all three bags they might say, "She didn't have one bag big enough" and card them all together.

When you get this carded wool back, it will be three separate shades. They may not be as different from each other as you expected, but they will be different enough for the desired result.

Now, to make a more interesting yarn than any of these

separate shades, you can just peel off layers of two or all three of the shades and tear strips off them to spin together as a subtly variegated yarn. With the soft-yarn method these will be somewhat unevenly variegated, which is good because the yarn then looks much more handcarded, and certainly more handspun. This is the real challenge of custom carding: how to keep it looking like handspun, like a crafted yarn as distinct from a manufactured product.

Spinning for
Softness &
Speed

4 Wheel Requirements and Limitations

For successful spinning by the soft-yarn method, there are certain wheel requirements. Your wheel must operate smoothly, drawing the yarn onto the bobbin as quickly as needed, or it will add more twist; it will also slow you down considerably. No matter how fast and deftly you are able to draw out the yarn, if you have to wait for the wheel to pull it in, you will lose the momentum that you need for speed. I admit to being an impatient person, and nothing tries my patience more than a wheel that will not snap in the yarn at my bidding.

Quite a few factors may be to blame when a wheel will not pull in quickly. Check these:

1. Too shallow a pulley ratio (on a double-belt wheel).
2. Brake band too loose (on a single-belt wheel) or slipping.
3. Poorly fitting bobbin (bobbin may be too tight, lack bearings, or need oiling).
4. Buildup of dirt or old oil in spindle inside bobbin.
5. Poorly fitted front or back maidens holding the flyer.

6. Orifice or hooks too small for size of yarn being spun.
7. Insufficient drive-belt traction (diameter may be too small).
8. Too loose drive-belt tension.
9. Fibers wound around spindle between bobbin and flyer.
10. Buildup of stray fibers wrapped around hooks of flyer.

SPEED

Although mastering the one-handed method can increase your spinning speed by eliminating unnecessary hand movements and by giving ease in the handling of fibers, there are other things involved in the consideration of speed. Some of these have just been mentioned as points to check when a wheel will not draw in quickly, and some are not at all related to that, only to spinning speed in general.

DRIVE RATIO

The primary limit on speed is the potential speed of the wheel used, and the chief determination of this is the drive ratio, that is, the ratio between the diameter of the drive wheel and that of the driving whorl. In double-belt wheels such as the Paul Dixon and Columbine and Haldane wheels, this would be the drive wheel diameter divided by the flyer pulley diameter. Here the speed of the flyer determines the speed with which the twist is introduced. The ratio is also the drive wheel diameter divided by the flyer pulley diameter in single belts having the brake on the bobbin, such as the Poly, Ashford, Nagy, and Lendrum. The drive ratio on single-belt wheels having the brake on the flyer, such as Louët, Alden Amos, and most of the bulk spinners, would be the drive wheel diameter divided by the bobbin whorl diameter. I am easily confused when it comes to figuring drive ratio, especially on double-belt wheels.

It is necessary to remember that "fast" wheels or "high" drive ratios are relative terms. A wheel that would be considered "fast" for a medium-heavy yarn with a low twist (such as 3 tpi) would be considered quite a slow wheel for spinning finer yarn (such as 10 tpi). Because of this, a wheel that can have its drive ratio changed at will, by the use of different bobbins, pulleys, or whorls, is most useful for a variety of yarn sizes.

Double treadle wheel with several pulley ratios. By Schacht Spindle Co. Inc.
6101 Ben Place, Boulder, CO 80301, 303-442-3212

DRIVE EFFICIENCY

The efficiency of a spinnning wheel — within its design limitations — is determined mainly by the absence of drive band slippage. Drive wheels with grooves for the belt will give better traction, but the grooves must not have a polished surface and the belt size should be suited to the groove size.

The drive belt diameter is one very controllable factor in the efficiency of the bobbin and flyer pulleys. For good traction, more than one surface of the band should come into contact with the pulley grooves. The groove should touch the band on three sides with a U-shaped pulley, and wedge it on two sides with a V-shaped pulley. This happens when the belt is the correct size for the pulleys.

To make the yarn draw in well, the amount of tension on the drive band must also be proper for the size and twist of the yarn being spun. This tension may have to be increased to a point

This pulley ratio is a good 2:1, but the drive belt should be larger in diameter to be efficient in these pulley grooves.

where treadling is difficult if the drive band diameter is not large enough in relation to the pulley grooves on the flyer and the bobbin.

PULLEY SHAPES

In addition to the proper belt diameter, there is one more variable that affects the traction. The shapes of the flyer pulley and the bobbin pulley make a subtle difference in the slippage and the traction.

The normal shape of pulleys for spinning a medium-weight yarn on a double-belt wheel would be a V-shape on the flyer pulley and a U-shape on the bobbin pulley. For the most efficient and fast spinning of fine yarn, these pulley shapes could be reversed, with the U-shape on the flyer pulley and the V-shape on the bobbin pulley. If the belt diameter is right for the size of the grooves, this will still give good traction on the drive wheel, and the V-shape of the bobbin pulley will cause the yarn to *start* to pull in faster as it is released to draw in. In fast spinning, a bobbin with a U-shaped pulley seems to take just an instant (which seems like a long time) to change from the bobbin slippage (needed when it is not winding on) to the positive traction that pulls in the yarn when you release it to be wound on.

Large orifice and auxiliary spinning loop on the little traveling Louët.

DRAW-IN RATE

The pulley ratio (on double-belt wheels) or the brake device (on single-belt wheels) will be critical to speed, because these control the rate of drawing-in of the yarn. If a double-belt wheel does not have a high enough ratio between the flyer pulley and the bobbin pulley for the size of yarn being spun, it will not pull the yarn in properly. This could be remedied by having a larger pulley made for the flyer or by having smaller grooves in the bobbins. You can compensate some for a low ratio by a tighter drive band, but beyond a certain point this will make it increasingly difficult to treadle.

If a single-belt wheel has a brake that cannot be adjusted easily and accurately, or has a brake peg that is prone to slip as you spin, it is an impediment to speed. The most common problem in adjusting these wheels comes from overadjusting, and a clumsy adjustment control will compound the difficulty.

BOBBIN LENGTH

A long bobbin permits more yarn to be spun onto it before it feels as though the brake needs tightening, while a shorter and wider bobbin might hold the same amount of yarn but would

Speed hooks are best for containing the yarn during fast spinning. Paul Dixon wheel.

Poly wheel from Pipy Craft, with a hook instead of an orifice for spinning highly textured yarns.

need more brake or drive belt tightening as it fills. It is most efficient to start filling at the end of the bobbin nearer you rather than at the opposite end, especially if the flyer arms taper out.

Most spinners consider a bobbin about five inches long to be a good generous size for production work. For fast spinning of fine yarn it helps to have speed hooks (of a slightly corkscrew shape) or a movable eye on the flyer. Otherwise the yarn tends to lift up

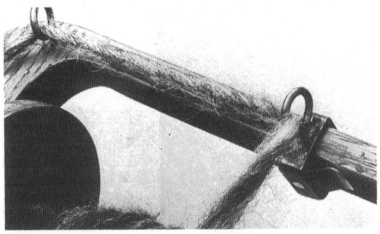

Sliding eye on Ernest Mason's Swiss-style wheel. The spring beneath the flyer arm allows the eye, which is secured by small pins in top of the flyer arm, to be moved from place to place.

out of the hooks as it shoots onto the bobbin at high speed, particularly at the far end of the bobbin.

THREADING HOOK

Spinning speed can be slowed down slightly by lack of ease in threading of the bobbin, and the need to use a hook to get the yarn through the orifice. A few wheels have a loop or hook instead of the traditional orifice, and some wheels have an orifice of a size and shape that does not require a hook for threading.

FLYER HOOKS

Yarn snagged on the tips of rough flyer hooks can require time to disengage. This snagging also causes a buildup of wool around the hooks, which must be cleaned off periodically. Smooth hooks can prevent this problem, as can a movable eye that you can slide along the flyer.

LEVEL-WIND

One aid to speed would be something that would eliminate the need to stop spinning in order to change the yarn from hook to hook on the flyer so that it winds evenly onto the bobbin. This is the level-wind idea, and has been invented and reinvented many times. One form of it, a system of moving the flyer back and forth, is found among the drawings of Leonardo da Vinci. John Antis later invented a level-wind mechanism which moved the

WooLee Winder, a level-wind device made by Robert Lee and Son, P.O. Box 941, Oregon City, Ore. 97045

bobbin instead of the flyer. The difficulty in these has been that the added parts have been too cumbersome, fragile, and complicated. Moving the yarn by hand kept the process simple and, in the long run, probably saved time.

Now there is a different type of level-wind for spinning wheels, similar to the device on fishing reels that enables the line to wind on evenly and automatically. The WooLee Winder was produced by Robert Lee and Son, to be used on the two styles of wheels built by Ernest Mason. It is now available for the Louët and other wheels. Its movable eye travels along the flyer arm automatically, activated by the movement of the bobbin each time the yarn winds onto it.

Using one-handed spinning and the level-wind device, you need only have all your wool within reach in order to spin a whole skein without looking at the bobbin or the yarn. The bobbin fills evenly, and when the yarn starts to touch on the flyer arm you know that the bobbin is full. You can either remove the bobbin for plying (if you ply) or wind off the yarn into a skein for washing and blocking.

FIBER PLACEMENT

Time can also be saved by the closeness of the additional fiber supply and the ease of reaching for it. While it rests the body to get up and move about at the completion of each skein, the wool to do each skein should be kept within reach to avoid wasted movement.

TREADLING

Ease of treadling is important for both speed and stamina. A weighted and balanced drive wheel can add to momentum. A double-action treadle (heel and toe action), especially one that allows the spinner to sit in a straight-on position in relation to the spinning wheel, has several points in its favor. It has a smoother action than most Saxony-type single-action treadles, requires less effort to operate, and is less susceptible to drive wheel reversal with beginners. It gives more control over the action of the drive wheel, permitting control on either forward or backward motion of the wheel. This means that it can be more easily set in motion from almost any wheel position. A double treadle (on which the spinner uses two feet, working treadles alternately), if properly designed, can also be effortless and give about the same advantages.

One reason a beginner should have what is considered a "slow wheel," with a low drive ratio, is that it permits treadling at a rate conducive to good momentum. With a faster wheel, a beginner would have to treadle much more slowly to achieve the same spinning speed. Too slow a treadling speed or erratic treadling can result in reversal of the drive wheel direction, throwing the yarn off the hooks. Single-belt wheels with the brake on the flyer have the potential for either slow or faster drive ratio, depending on the size of the bobbin pulley in relation to the drive wheel diameter. Since a different bobbin with a smaller pulley groove would give a faster drive ratio, this change would speed up the wheel when the spinner became more efficient.

There are two different systems of treadling. One system is for the feet to treadle at a constant pace and the hand action to endeavor to keep up with them. In the other method, the feet and the hands keep pace with each other, working in a coordinated manner suited to the problems of the yarn production. One-handed spinning works well with either way of treadling.

Saxony treadle.

Double-action treadle.

Double treadle.

TYPES OF SPINNING WHEELS

It is generally agreed that there is no "best" wheel for everyone. There are well over fifty spinning wheels of many different styles and capabilities currently available, which can be evaluated comparatively only if you have all the pertinent statistics on each. While there are subtle qualities that do not show up in measurements and ratios, the statistics are clues to the theoretical potential of each wheel.

What follows here are general observations on the different types of wheels, related to their capability for soft yarn production.

Double-belt Wheels

I find that most double-belt wheels work well for one-handed spinning when they have a 2:1 pulley ratio, meaning that the flyer pulley is twice the diameter of the bobbin pulley.

This is a much higher ratio than would ordinarily be suggested for a medium-fine yarn, but if you use a belt tension much looser than normal your wheel will draw in firmly yet not tug too hard on the yarn as it is being spun. When spinning fine yarn, you want enough belt tension to get the traction needed to turn the drive wheel and the pulleys, but not a lot more. It is hard to give any exact ratios for this, because performance varies from wheel to wheel, but the ratio of 2:1 is a good starting point.

Of course, pulley ratio is still related to the size of the yarn being spun. If you have only a shallow pulley ratio it should still be adequate for a much finer yarn; the shallower the pulley ratio, the finer the yarn. And, since the one-handed method is especially good for fine yarn, you need only determine what size yarn your wheel will handle best.

With a double-belt wheel you need a sufficient pulley ratio, correct belt tension, and a heavy enough belt for good traction in the pulley grooves.

Single-belt Wheels

The brake adjustment is important for one-handed spinning with single-belt wheels. Whether the brake is on the bobbin or on the flyer, the degree of tension on that brake still controls how well it pulls in.

On this wheel, the large orifice does not need a threading hook, the brake is easily adjusted and will not slip, and drive ratios can be changed by reversing the bobbin. (Wheel from Louët Sales Spinning Wheel Co., Box 70, Carleton Place, Ontario.)

On a standard-size flyer wheel with a single belt — such as the Louët, with its self-adjusting drive belt and brake on the flyer — either soft fine yarn or soft medium-weight yarn can be spun. The tension on the brake is the only adjustment to be made, and this wheel is one of several that have what is called a precision brake. Its screw device stays where you put it and cannot slip while you are spinning. You can tighten it a measured amount, such as a half turn or a whole turn (if needed) while the bobbin is getting full. When you have emptied the bobbin you can then turn it back again, accurately, to the same tension it had at the beginning of the skein. This feature is also found on all the Alden Amos wheels. It makes it much easier to keep the same size yarn from skein to skein of a project, where uniformity of the skeins is important.

The Ashford wheel has the brake on the bobbin instead of the flyer, and there are two separate adjustments to be made: the drive belt tension and the brake tension. While the brake tension is more crucial, they will both need to be right for efficient spinning. Adjust these a very little at a time, the drive belt first and then the brake. As you get a bobbin about half full, it usually needs the brake tightened more to keep it pulling in well.

Single-belt Bulk Spinners

Bulk or "Indian head" spinners are especially designed for spinning of heavy yarn. They have an eight-to-ten-inch bobbin, a brake on the flyer, an orifice of about one inch, and often a leather drive band. Even with the brake off, they usually pull too hard for the spinning of fine yarn, and especially of fine soft yarn.

While the large orifice permits the spinning of thick yarn, it allows a fine yarn to wobble as it rotates in the orifice. Instead of the yarn being drawn smoothly onto the bobbin, there will be a discernible jerk; this can be annoying, and may also produce unplanned thick and thin places. The wobbling can be offset somewhat by holding the yarn off to the side as you spin, thus pressing it against one side of the orifice. The problem can also be remedied in part by having a plug with a smaller opening made for the orifice. Still, the drive ratio is low, in keeping with its intended use to spin heavy yarn, and it would be very slow for the spinning of finer yarn.

It would certainly be difficult—in fact, nearly impossible —to *learn* to spin a one-handed soft yarn, and particularly a soft fine yarn, on a bulk spinner rather than on a more standard model of wheel. Soft heavy yarn can be spun on it with a short draw, or by tearing carded wool into strips and spinning them on by the twisting action without any further attenuation of the wool. A combination of the short draw and an unsupported long draw will produce a heavy yarn with a good amount of texture.

Great Wheels (Walking Wheels)

With the great wheel the yarn has the *exact* twist you put into it when you decide to wind it onto the spindle (as opposed to flyer wheels, which do add some additional twist as the yarn is drawn onto the bobbin). The great wheel is excellent for spinning short fibers and fine yarn. A comprehensive and interesting book on this is *The Legacy of the Great Wheel* by Katy Turner (Select Books, 1980).

Anyone intending to eventually use a great wheel should remember that most great wheels—although not all—are set up for left-handed draw only, with the right hand used to turn the wheel. With this in mind, you may want to learn to draw

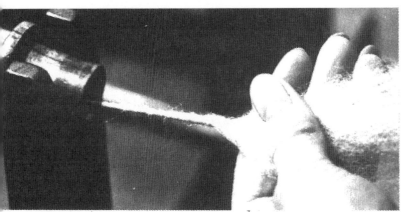

The extra-large orifice of a bulk spinner causes a wobble when finer yarn is spun.

with your left hand, or learn equally well with both hands. With the great wheel one-handed spinning is not optional, but necessary, since the other hand turns the wheel. What *is* optional is the exact posture of the spinning hand, and even this is somewhat dictated by the way the wool has been processed.

The palm-held fiber position, useful with drum-carded and mill-carded wool in particular, gives the main advantage that you do not have to look constantly at the hand that is doing the spinning. In order to learn this technique on the great wheel, it is best to start out by holding your hand with palm up in order to first learn the feel of the wool being pulled and twisted as your hand draws away from the spindle. Later, in actual spinning (as opposed to practicing) on the great wheel, this hand position is altered somewhat: not palm down, but halfway in between. With your palm already familiar with the feel of the fiber slippage, the use of the hand in this sideways position will seem quite natural. In fact, this may be the position you will assume with any wheel, once your hand is adequately trained. The wool may even be touching your thumb and index finger, but it will be an even lighter touch than that of the two fingers that hold it against your palm. The thumb and index finger will be resting very lightly against the wool, ready to tighten down on it if the fibers in your palm start drawing too thin.

Treadle Spike Spinners

A treadle spike spinning wheel does not necessitate a one-handed spinning method in the same way that a great wheel

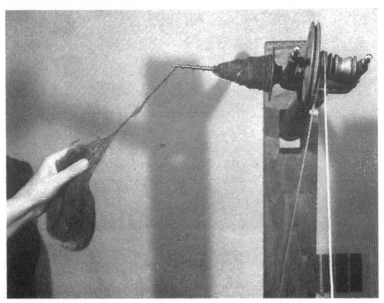

Proper hand position for spinning on the great wheel. Photo from The Legacy of the Great Wheel *by Katy Turner, Select Books, Mountain View, Missouri, 1980.*

does. Because of the treadle, both hands are free; therefore both hands can be used to draw out the yarn.

If a treadle spike wheel is designed like a castle wheel — as is the Penguin Quill, for example — either a right-hand or a left-hand draw may be used, since the spike will be pointed directly at the person who is spinning. The spinner, seated at the wheel, can use one hand or both, and can draw out in either direction. However, the yarn cannot be drawn out as far as on a walking wheel.

On a treadle spike wheel made like a great wheel — such as the Rio Grande — the spike is at a right angle to the spinner, who must draw out in the direction the spindle is pointed. There is still the option of using one hand or both.

Treadle spike spinners are ideal for heavy or high textured yarns because the yarn does not have to go through an orifice. Even overtwisted yarn is no problem to wind on.

Antique Wheels

Antique wheels (and most "replica" wheels) have not only a small orifice and small hooks, but a very shallow pulley ratio (ratio between the size of the flyer pulley and the size of the bobbin pulley). They were intended for the spinning of very fine

43   Wheel Requirements and Limitations

yarn with a fairly hard twist, for serviceable apparel. They are not ideal for a long sweep with one hand or for the long draw, or for any method that requires a whole length of yarn to be pulled in quickly. They are more suited to the short draw, or a shorter version of one-handed spinning.

While the extreme gentleness of the pull would make it difficult to learn the soft-yarn technique on an antique wheel, once the technique is learned it should no longer present much of a problem, as long as the size of the yarn being spun is coordinated with the capability of the wheel's pulleys. Since the present method is the fastest way of spinning fine yarn, and fine yarn is the kind the wheel is most able to handle, it could be worth the struggle. Once learned, this method might help control the excessive overtwist that is so often seen with antique wheels.

If the wheel does not want to wind on the size of yarn that otherwise could be spun with the orifice and hook size, check through the list of other factors at the beginning of this chapter. If none of these applies, then consider having a larger pulley made for the flyer. Another possibility, to achieve the same result, is to have the bobbin pulley made smaller on a lathe, unless it is already too small to have more wood removed.

Without a slightly larger pulley ratio than was intended for the tightly spun yarn which it was designed to spin, it is difficult to get an antique wheel to draw in fast enough and firmly enough so that a really soft yarn can be spun on it. The exception to this is an antique great wheel, which of course has a spindle action instead of a flyer action. If in good condition, it will be admirably suited for the production of soft fine yarn.

Opposite page: *Unsupported long draw on the Rio Grande Wheel (Wheel from Weaving Southwest, 216-B Paseo del Pueblo Norte, Taos, NM 87571.)*

Spinning for
Softness &
Speed

5 The Compleat Spinner

As I indicated in the last chapter, one single technique will not really suffice to handle the distinct requirements of various types of wool.

While the soft-yarn method should be *learned* as a single technique, it will actually be used in combination with others. Barring bursitis or a broken arm, you will have little occasion to spin a whole skein with one hand in your lap. What you should aim for is the *minimum* use of the other hand, which should be used only when it is needed.

Techniques with which the one-handed method combines especially well are those called the long draw and the unsupported long draw. (See Appendix for a more detailed description of these two techniques.)

THE LONG DRAW

In the spinning of finer yarns you will have less occasion to use the long draw, inasmuch as any threat of overtwist can be remedied by drawing back faster with the hand that is drafting the wool.

With medium-weight yarns, it is more practical to use both hands whenever overtwist or other problems arise. If too much twist builds up—which, unless corrected, could stop the drawing out of the fibers

46

entirely — the alternate use of the long draw is one way of handling the situation. Use your free hand to grasp the yarn near the orifice and pinch off the twist. The hand that holds the fiber supply moves back along the fiber mass (away from the orifice) and holds on more firmly, drawing out that whole span to spread the twist over a longer section of yarn. This must be accomplished quickly, so that you can release the pinched-off twist before enough accumulates to add overtwist to the section you have drawn out.

THE UNSUPPORTED LONG DRAW

A similar action can be used if an unexpected or undesired slub should occur. To even it out with the unsupported long draw, the hand that is holding the fibers will have to move quickly and grasp the fiber supply further back. Your free hand pinches off the twist near the orifice, with a staccato pinching action, letting the twist come through as needed. You then use the unsupported long draw to "stretch out" the area into which the twist is entering and thus attenuate the unwanted slub. The slub, being the part with the least twist, is the part that will be affected. If you do not move fast enough, or if you allow too much twist into the slub as it is being stretched, this will limit how well it can be drawn out. As you reach the point where there is no more stretch in that section of yarn, it must be wound onto the bobbin immediately.

This does not work at all for a slub in overtwisted yarn, because the overtwist usually locks the slub in quite firmly. Neither does it work very well with uncarded wool, which does not have uniform slippage. Both of these cases would require the use of both hands, but used close to the slub, to draw it thinner. This is not ordinarily a problem when using uncarded wool, inasmuch as you are probably spinning it for maximum irregularity.

While slubs are usually blamed on inadequate wool preparation or inept spinning, most often you will find that a small seed or other foreign matter is the nucleus of the slub and has prevented the free slippage of fibers.

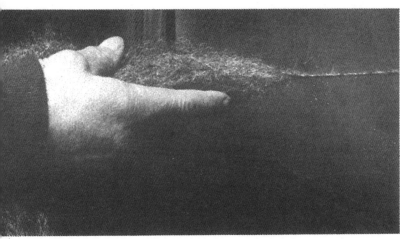

Wool drawing out of a larger mass in front of the hand.

YARN SIZE

Part of your control over the size of yarn you want to spin will be in adjusting the brake (for single-belt wheels) or the pulley size and drive belt tension (for double-belt wheels). The rest of the control will be in the use of your hands.

For fine yarn you need an especially light touch, and you will need to retreat quickly ahead of the twist being inserted by the spinning wheel action. For medium-weight yarn you may need to use your thumb more, and exercise a firmer control over the wool supply.

For heavier yarn, you may still be able to spin with one hand, with the yarn drawing out from a larger mass, further in front of your hand. This is still effortless, but you will have to look at the yarn. With heavy yarns it is usually more efficient to use your other hand as needed, rather than trying to do it all one-handed.

FANNING OUT

A technique seen in some parts of the country is fanning out of the fibers while spinning. There are several ways of fanning out. All have the effect of opening up the fibers more and widening the triangle of the fibers where the fiber supply meets the twist.

1. With a great wheel (walking wheel), either carded wool or combed locks of fleece can be fanned out, usually over the heel of the thumb, and spun with one hand.
2. Using both hands, mill-combed top or combed fleece locks can be fanned out over the heel of one thumb and

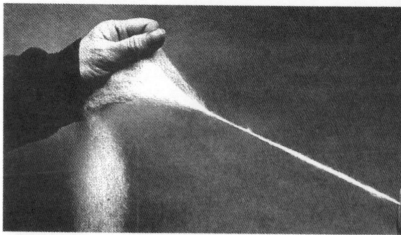

Method #1: fibers fanned over the heel of the thumb.

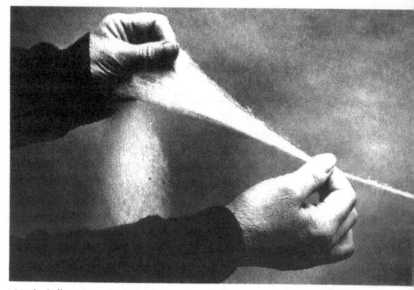

Method #2: fibers fanned with one hand, drawn out by the other hand.

drawn out into a worsted yarn by the other hand, which is closer to the orifice.

3. Carded or uncarded wool can be fanned by using both hands, the wool held between the hands and fanned out almost as in teasing.

The first of these three ways is an excellent technique, particularly for the great wheel, as it assures that you will not encounter fibers that cannot be drawn out with one hand.

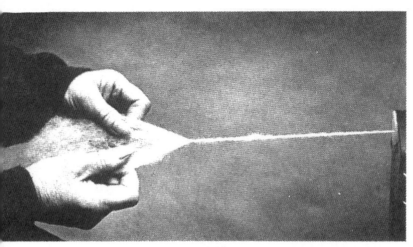

Method #3: fibers fanned with both hands.

The second way would go faster if done with one hand, unless the wool is gummy. To obtain the parallel arrangement of the fibers necessary for spinning worsted yarn, much hand wool combing is done with grease wool. This is sometimes so sticky that both hands are needed in order to produce a smoothly spun worsted yarn.

The third way is fascinating to watch, but does not lend itself to fast spinning and does require close attention. It too is a way of dealing with wool that is either too gummy or too dry to draw out evenly with one hand. It does work exceptionally well with unwashed and uncarded wool of an open type of fleece that is not too greasy.

In soft spinning with one hand, a brief one-handed fanning out can be useful when necessary to open up problem areas in otherwise well-prepared wool.

HAIRINESS

Sometimes you may want a hairy yarn. But when you want to prevent hairiness, your free hand can be very useful. Lightly smoothing the yarn as it is being drawn onto the bobbin will reduce hairiness by catching more fibers into the twist. Also, grabbing at any stray fiber that appears to have escaped the twist will minimize hairiness. Try this at slow speed and see how detaining this fiber, for just a second, can cause it to be completely caught into the twisting action. This hand movement can be done so smoothly and quickly that it will not require slowing down.

Preventing hairiness by detaining stray hair with the free hand.

In fast spinning, these actions cannot be done slowly and deliberately; they must be accomplished quickly so as not to disrupt your momentum. But this is only possible if with one hand you are in full control of the wool, enabling your free hand to act quickly and independently wherever its services are needed.

THE FREE HAND

Once you get in the habit of using the free hand only where it is actually needed, the way in which you use it will be determined by the kind of yarn you are spinning and the wool which is being spun. With long wools you can cultivate a relaxing rhythm with the long sweep ahead of the twist, then a momentary assistance with that free hand to add a finishing touch to the desired yarn size. Then let it pull quickly onto the bobbin.

Poorly prepared wools and long wools may both require the use of the free hand more often than medium-short fibers and those that are well carded. With medium-weight yarn, the hand will be used to provide a solid firmness to pull against in the last stage of almost every draft. This last brief pull is an important factor in determining the character of the yarn, but may not be needed if you want the maximum natural irregularity. I use the term "natural" in contrast to a contrived or deliberately placed slub. I could have said "accidental," except that there is, with every wool, a certain feel, a certain predisposition to the formation of a unique and one-of-a-kind yarn. This can be encouraged and brought out by using a light touch so that the natural tendency is allowed to emerge.

Even difficult wools, needing a heavy hand, have their own unique characteristics that can be encouraged. If you coordinate your way of spinning with what you perceive to be the inherent nature of the particular wool, you can achieve a great many varied and even exciting effects.

Spinning for
Softness &
Speed

6 Teaching the One-Handed Method

O ne-handed spinning is easier to teach to beginners than to advanced spinners who are already comfortable with their own method of spinning. It helps if you explain the advantages, for when students understand the reasons for learning a different method, they are more likely to make a greater effort. Point out that the one-handed method is a problem-solving technique enabling the fast and effortless spinning of soft yarn, bringing freedom from mannerisms that can only hinder spinners in their efforts to gain greater skill. It also means the end of overtwist.

INCHING

Overtwist is customarily explained by saying that the feet are moving too fast for the hands. "Treadle slower" is the usual advice. It is true that slower treadling (and/ or a slower wheel) will allow the spinner's hands more time to distribute the twist into yarn. However, if a spinner is inching along in the way people generally do when first learning to spin, habits are already being established that can be hard to break. This is not to say that a beginner should not be started out in this "inchworm" manner, but I would liken inching to the dogpaddle in swimming, by which the swimmer gains enough confidence to attempt strokes that require a more disciplined use of hands and feet.

INTRODUCING NEW TECHNIQUES

Spinning, by becoming more disciplined, also becomes more effortless and more efficient. Having discovered that they can spin, students are ready to be taught other, more efficient, ways.

After the first lesson or two, I would suggest that students go directly into the one-handed style. What can be taught is that the spinning wheel can now take over the function of the hand that is nearer the orifice. Instead of being drawn out by the hand, the fibers can be drawn out by the twisting and pulling action of the spinning wheel.

Students can also be introduced, about this time, to the short draw and the long draw, and then to the unsupported long draw. It has been my experience that the sooner spinners are acquainted with all methods, the easier they will find any one single method. It is only by being comfortable with many ways of spinning that a spinner can evolve a style both personal and highly productive.

The student will also be challenged by the presentation of separate but complementary techniques. As we say at our house, "There is more than one way to spin a batt."

DEMONSTRATING

To teach the one-handed method with any aplomb, an instructor should be confident of being able to do it with almost any type of wool and with almost any spinning wheel. As was mentioned in Chapter 4, with a bulk spinner or with most antique wheels it is really difficult for a student to learn this technique. It is not impossible, but it is certainly being done under difficulty. A spinner who has either type of wheel should be encouraged to borrow a different wheel just long enough to learn this method before trying it on her own wheel.

It is easier to demonstrate the soft-yarn technique than to describe it, but repeated demonstrations will be necessary (after the students have tried to do it), as will a description of how the process feels when it is being done correctly.

Showing a student how lightly the wool is held in the palm.
Photographed at Ram Wools workshop in Winnipeg.

TESTING THE WHEEL

Before starting to help an individual student, test the wheel to make sure it is correctly adjusted. In order to teach the method successfully, you will need to convince each spinner that it truly can be done—and done on *her* wheel.

Have the student do the treadling while you spin. The hardest thing for you to get used to, with a class full of strange wheels, is the different treadle action of each. Your first efforts on another person's wheel may not be very impressive if you have to quickly adjust to a strange treadle action while at the same time convincing the student that the method you are demonstrating is an easy one.

ASSISTING THE HAND

Next, help the student to do the spinning. The best way is for the student to treadle her own wheel while you hold the hand that controls the fiber supply. This may sound like asking one person to hold the needle while another person threads it, but it really works.

Have the student hold out the hand she normally uses to hold the fibers (the hand further from the orifice.) This hand should be held out *flat*, not hanging onto the fibers at all. Press the fibers against her palm very lightly as she treadles, so she gets the feel of the fibers slipping against each other in her palm as the yarn is drawing out. You will be guiding her hand also, showing her how fast to draw out in order to stay ahead of the twist produced by the spinning wheel, and when to allow the yarn to wind quickly onto the bobbin. You must also move her hand back at the proper speed in order to keep a light tension on the spun fibers so they do not wind on until desired. Your hand on her hand dramatizes the ease and simplicity of this method and vividly focuses attention on the need to feel the slippage of the fiber as the wheel draws it from her hand. If she cannot feel this movement, have her close her eyes; this heightens her awareness of the action of the fibers, which is the sensation that must be cultivated. Once students recognize that feeling, they know what they are trying for, and they also know how lightly the wool must be held. Do not encourage them to spin too slowly, for at slow speed there is not sufficient movement of the fibers against the palm, making it more difficult to feel the action.

THE LIGHT TOUCH

When helping a student by holding the wool as she treadles, the teacher will chiefly be trying to keep her from holding onto the fibers. The student's inclination (I could call it a compulsion) will be to hold onto the wool, no matter how plainly she is told, "Just hold your hand out flat while I hold the wool."

The student must experience the feeling of the fibers moving as they are drawn out by the twist, and of how lightly the fibers are actually held so that they are free to form into yarn. Even after this demonstration, it is almost impossible to convince students that the fibers *can* be held so lightly. The teacher will still have to go from one student to another, prying their fingers off the wool when they hold it too tightly, and trying to keep them from using their thumbs. (One girl, in desperation, put a clothespin on her thumb!)

Students will say, "It will get away if I don't hold it tighter," and sometimes it does. But if a previous problem has been excessive twist, and then if from undertwist the yarn pulls apart, that will be some evidence that they are beginning to conquer their overtwist. With practice, they will sense the amount of twist that is most suitable for the particular wool they are using, and the purpose for which it is being spun. It is that sensitivity to the nature of the specific wool that is essential to fast and proficient spinning.

HAND MOVEMENTS

If it is taught to beginners before they have been spinning for any length of time, the soft-yarn method will be much easier for them than it will be for the experienced spinner. Having little to unlearn, beginners do have an advantage.

No matter how well and how easily advanced spinners catch onto this technique, their hands may revert to their customary spinning habits at their first lapse of attention. However, when they have been using it for a while, they will find that their hands will do spontaneously whatever the particular wool requires. They will not have to consciously decide what technique or combination of techniques to use. Well-trained hands, if they have a variety of methods to draw from, will make the right decision.

*Spinning for
Softness &
Speed*

7 Projects for Soft Yarn

S everal practical questions come up in regard to the uses of softly spun yarn.

PLYING

Can you ply your soft yarn? Of course. However, you won't *need* to ply it in order to make it soft (which is one of the reasons that some spinners ply their yarn). If you like the *effect* of plied yarn, or the look of two or more colors plied together, then there is no reason that you should not ply. If you are spinning yarn for sale, then the economics of the situation may guide your decision.

KNITTING

Can you knit with soft single ply? Doesn't it have to be plied to keep it from skewing off on the diagonal when knitted?

It has been our experience that only the *harder twisted* handspun singles will go off on a slant when knitted. Everything that I knit is done with softly spun single ply, washed and blocked, and this problem has not occurred.

FUZZING AND PILLING

There is a more undeniable problem with the soft spun singles: they do fuzz up more, and pill. However, there are several ways to remove the surface pilling.

Notion counters sell a small D-Fuzz-It comb that successfully removes these surface pills. It doesn't look like a comb, but is a golden mesh edge on a brown plastic handle. The directions say, "Just hold fabric taut, brush quickly in one direction." If this is not available locally, it can be ordered from Susan's Fiber Shop, N 250 Hwy. A, Columbus, WI. 53925.

Some people use manicure scissors, or an electric razor, or, on knitted surfaces, even fine sandpaper.

A few spinners "shock" their yarn after spinning, and find this useful in reducing the pilling as well as in preventing both stretching and shrinkage of the yarn. The shocking process consists simply of washing the yarn in very hot water, then rinsing it in hot water and immediately dipping it in cold water. This felts the fibers of the yarn.

Other spinners brush their sweater or fabric with a hairbrush or a metal lint brush to make a napped surface, and repeat the brushing from time to time.

WEAVING WARP

In her book *Spin Span Spun*, Bette Hochberg points out that mill-spun yarn has been around for barely 200 years. With spinning at least 10,000 years old, weavers had nothing but handspun for ninety-some centuries. That means warp, as well as weft, was handspun.

But can you use soft one-ply handspun for warp? Yes. I happen to like a soft warp, so this is all that I use. It does need warp sizing to make it a guaranteed success. (See Appendix for sizing information.) On the other hand, if you enjoy working with a harder twist warp yarn, sizing may not be so necessary. I would use sizing on all my warps, even though some may not appear to need it. The loom-shaping that I do on my jackets and vests would be a strain on the selvage threads, even for a hard-twist yarn, and the sizing minimizes wear and tear on the yarn during both warping and weaving. It washes out easily. I do not use a different twist yarn for selvages, or even double-sley them.

I had been spinning for about five years when I read that "you cannot use a one-ply handspun for warp" — but I had been doing it for almost five years, so I was told too late. Even *Wool Gatherings* (Hamilton Wool and Craft Guild, Victoria, Australia) says that single-ply handspun wefts are satisfactory, but that "a single ply handspun warp is *not* satisfactory." Now, when anyone says "You cannot . . .", I usually think, "Perhaps *they* cannot but someone who really wants to may find a way to make it work."

There are ten commandments, but none of them have a thing to do with wool and spinning.

This is how I have it figured about using soft single ply for warp: if you can make it into a skein and wash and block the yarn and it holds together; if you can wind it into a warp chain and it still holds together; then warp-sizing the chain will make it suitable for warp use in garment weaving.

Most weavers prefer string heddles instead of metal ones for weaving with handspun. I have used metal heddles, but feel that string heddles are much kinder to the warp, so all of my looms have string heddles.

In order to eliminate loom waste, I use a "dummy" or permanent warp, which makes it possible to weave right to the end of the handspun warp. (See Appendix for dummy warp directions.)

WEAVING PROJECT:
THE SHEPHERD JACKET OR COAT

The shepherd jacket, shown at the end of this section, is an ideal project to be woven with softly spun yarn. Very little cutting and tailoring is required, and there is a minimum of tedious loom-shaping. The basic design has been around for a very long time, but the neck detail is original. It can be woven as a short jacket, or fingertip length, or coat length.

Because of its loom-shaped neckline construction, it is more wearable than a jacket with a front edge extending straight up to the neck. Its three-quarter kimono sleeve looks good over anything with long narrow sleeves. Woven on a 45-inch loom, it fits sizes 8-10-12-14.

Before planning the weaving of this jacket, cut and fold a piece of paper to look like the jacket drawings. This will give you a clear idea of the parts of the jacket before and after tailoring. You will see that the outer edge of your warp should correspond in color to the center of the warp, if you are using warp stripes, because the center divides to form the upper

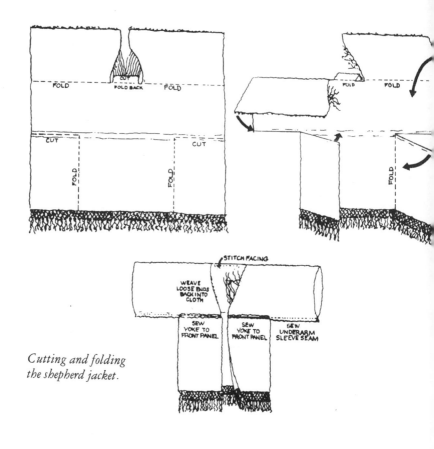

*Cutting and folding
the shepherd jacket.*

fronts, and these should present a continuous color along the
front edge of the jacket after tailoring.

Materials Needed

Loom:
45″ wide, using 5 epi in 5 dent reed, 224 thread warp (6 epi in 6 dent could be used instead).

Warp:
Handspun yarn of size about the diameter of a No. 2 to 4 knitting needle (or a round wooden cocktail toothpick). Warp sizing should be used on your chains (see Appendix for sizing instructions).

Weft:
Handspun yarn about the size of a No. 6 to 8 knitting needle (a little heavier than a wooden matchstick).

| Seam yarn: | Handspun about the same size as the warp yarn, and the same color as the weft. |
| Dummy warp: | By warping up with cotton carpet warp, then using it as a permanent warp to tie onto with your handspun warp, you will not have to allow for a lot of loom waste at the end of your jacket. You can weave almost up to the knots where the handspun warp ties onto the dummy warp. |

WEAVING

Bottom to Underarm: Tie your warp onto the front apron and weave in three header strips. You should have used, with knotted yarn and headers, about ten inches of warp, which will later be tied into triple-knotted fringe for the bottom of your jacket. I do my jackets in plain tabby weave, which is most practical for the size of yarn being used because it eliminates overshots. For a short (hip-length) jacket, weave 13 inches, patterning in natural or dyed yarns as desired. For fingertip length this would be 16 inches, or for coat length approximately 24 inches. This length will vary depending on the amount of elasticity in your yarn, estimated take-up and shrinkage, and height of wearer.

Underarm to Shoulder Line: This brings you up to the underarm. Here you actually have two choices: you can weave the finer "seam" yarn all the way across, in which case it constitutes a band of finer yarn across the back of the jacket, or you can use the seam yarn for the outer one-fourth of the width of the jacket, putting it in with a tapestry beater, two of seam yarn to one of regular weft. Directions here are for the first of these options.

Weave a 1/2-inch of tabby using finer seam yarn, which is about the size of your warp. Use a *very* large arc of weft, bubbling it a little to add more yarn, so that this finer yarn will not draw in the width of our weaving at that seam line. At the 1/2-inch distance, lay in two markers of bright-colored thread along with one of the weft picks. Each will be about one-fourth of the width of your fabric (about 10 3/4 inches) and extend from the outer selvages across the outer fourth of your fabric. These mark your two cutting lines for tailoring; the exact length will be adjusted later.

Weave another ½-inch of the fine seam yarn, still throwing a bubbled arc.

Now, change back to your heavier weft yarn. Weave another 13 inches. Even if you are making a longer jacket and have woven a longer distance up to the armhole, this 13-inch measurement from underarm to shoulder line remains the same. Stop with your shuttle at the right edge of the jacket.

Top Back of Jacket: Divide the width of your warp at the center. If you are doing several jackets, this is easier if you mark the center of the warp in the reed. Count off 13 threads each side of the center. You will be leaving this section of 26 center warp threads unwoven for a time, while weaving on the two side areas.

Using two shuttles, enter one shuttle from the right-hand selvage and bring it up through the top of the shed just *before* the 26 center threads. Start with the other shuttle and enter it in the same shed, from the top of the shed, just *past* the 26 center threads.

Using these two shuttles, weave on both side sections for 2 inches. Then, stop weaving on these side sections while you weave in the tab for the back of the neck. This tab gives firmness to the neckline as it is turned inward, and is a good place to sew in your woven label.

Tab: Using a quill of your seam yarn (warp size) and a small tapestry beater (or wide-toothed dog comb or table fork) weave back and forth across the center section, pressing in firmly against the woven portion. Weave a firm tab of about 1½ inches of tabby, trying *not* to narrow in the edges of the tab. If it draws in, it will put more strain on the threads that you will soon be picking up at the center edges.

End the tab with your yarn at the right side. Break off yarn, leaving about 10 inches of yarn, and use it to overcast the finished edge of the tab.

Slanted Neckline: Now continue on with the side sections, using two shuttles and your heavy weft yarn, and start shaping the neckline. This is done by incorporating the center threads, one at a time, back into the side sections. Catch one of the unwoven center warp threads back into the woven area each time the shuttle passes back from the center selvage.

To make a firm and neat selvage when catching in each warp

Shepherd jacket with warp and weft of handspun singles. This can be woven as a short or fingertip-length jacket or as a full-length coat.

thread, the weft should be pulled tight at the center selvage, even though you are still throwing a wide arc. One way to do this is to slant your weft out at an angle, and snug it up firmly at the selvage. Holding onto the yarn at the center selvage, close the shed on the slanted weft. Then, adjust this slanted weft to more of an arc (with the shed still closed) and beat it into place. This way you avoid small loops at that selvage, without drawing in the jacket width.

When all of the center warp threads are again incorporated into the weaving, the center selvages should be woven less firmly, in order to continue lessening the space between the center edges.

Jacket, woven but unsewn, washed and pinned onto a blanket to dry. Note cutting-line markers.

Neckline Down to Armhole (Fronts): Continue weaving with two shuttles and your warp divided in the center until you are 13 inches from the back of the neck (where you started weaving your tab).

Front Seam: Now, using your fine seam yarn with a bubbled arc, continue weaving on the divided warp threads for ½ inch, ending with the shuttles on the right-hand side of each piece. Break off yarn, leaving about 24 inches attached to each half of the fabric, and overcast each edge with it. Cut the jacket piece off the loom about ½ inch from the overcast edges.

Fringe: Untie your warp ends from the front apron and remove all but one of the header strips. Remove the last header strip just ahead of the knots as you tie them, and knot your fringe close to the woven piece, knotting four warp ends at a time. Make two more rows of knots, for triple-knotted fringe, taking two threads from adjoining knots for each knot in the the next row.

Make one row of machine stitching, with matching sewing thread, on top of the overcasting on the ends of your woven piece. Machine stitch around the three edges of the tab also. Leave the warp threads attached to the tab until after the piece is washed

Washing

Wash the jacket piece in hot detergent water. I let mine soak for about 15 minutes to half an hour. This is not just to remove the sizing, which washes out easily, but to help fluff up the yarn and tighten the weave. Rinse the jacket piece and spin out the water in the spin cycle of your washer, or roll it in towels to absorb excess water. If you use the spin cycle, be sure to remove the material the minute it is spun to avoid wrinkles.

Dry the woven piece on a blanket-covered table (or a clean carpet) laid out flat. By smoothing it well with your hands when placing it out to dry, you can usually eliminate the need for further pressing. Leave it until it is completely dry.

Tailoring

Trim off the excess 1/2 inch of warp threads from the overcast edges. Check and correct the length of the marker threads as follows: Measure off half of the distance across one of the divided ends of the piece (along overcast edge). This is the correct length for each of the markers. Shorten the markers, if necessary.

Stitch (zig-zag can substitute for two rows of stitching) around each marker thread before cutting. Starting at the selvage, sew close to the marker, up along one side of it and back on the other side to the selvage again. Do a second row of stitching, 1/16 inch further from the marker. Although this machine stitching will be inside a seam, it looks best if the thread matches the jacket color.

Cutting: Only two cuts need to be made for the tailoring. These are right along on top of each marker thread, inside the machine stitching which prevents its raveling. After cutting, I do one row of zig-zag stitch on the whole cut edge (both sides).

Seams: taking in the 1/2 inch of the woven fine yarn as a seam allowance, sew the underarm and across-the-front seam (this is one continuous seam as shown in the photo). The portion of the seam that is in the sleeve seam is not too crucial, but the part

Inside-out jacket showing sewn seams.

that is across the front of the coat must be sewn carefully, for a crooked seam is noticeable there. I sew it once with the loosest (basting) stitch of the machine, and if it is exactly right, I sew again over that same stitching line, with smaller stitches. If the first stitching was not correct, it is easily removed by pulling on the underneath (bobbin) thread, and can then be resewn.

Press seam allowances open on the inside of the jacket and sew them by hand against the jacket with matching thread. Make this stitching invisible by catching the sewing thread only into the *warp* threads of the front of the jacket. If you catch the weft threads, you will be able to see a shadow of this seam allowance on the right side.

Neckline: Leave the neckline warp threads attached to the jacket fronts, but cut them loose from the tab, cutting close to the overcast edge of the tab. Stitch the tab down, folding it inward at the very edge of the start of the tab weaving, and stitch close to that folded edge.

The threads that remain attached to the front edge of the jacket should now be worked back into each adjacent weft row. Taper (by fuzzing and tearing off, not cutting) the loose end of each hanging thread, leaving it about 1½ inches long, and weave it back into the adjoining row, beginning at the place where it is attached, with a large blunt yarn needle or a crochet

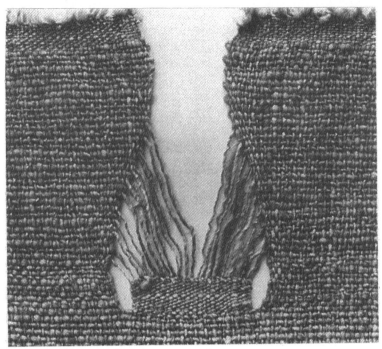

Detail of neckline and tab for woven label.

hook. This is done on the inside of the jacket, and if done carefully it will be absolutely undetectable.

Edging: To finish off the jacket, I use a simple single-crochet edging of the whole front. Using yarn about the weight of the warp yarn, work about six rows of single crochet, starting at the bottom edge of one front, working up the front, across the folded and stitched edge of the tab, and down the other front. Go back and forth to make about a 1 1/2-inch width of crocheted band. Tie in a false fringe at the bottom ends of the band, and untie the end knots of the fringe on the bottom of the jacket so you can work them together with the edge knots of the band fringe to make a continuous bottom fringe. Steam the false fringes over hot steam coming from a boiling tea kettle to fluff them up. (Steam from a steam iron is not sufficient.)

If the crocheted edging is done at a proper tension, very little steam pressing will be required for the edge. To do any steaming on it, use a wet cloth under an iron set at Wool temperature. Sew your label onto the neck tab, and the jacket is completed.

I wrote this up first in the magazine *SpinOff 77.*

KNITTING PROJECT: A SLEEVELESS SWEATER

This man's sleeveless pullover sweater is an easy one to knit; it has very little fancy shaping to do, and a minimum number of seams to sew when done. It is equally suitable for a conservatively spun yarn or for a novelty yarn spun from uncarded wool. Whatever kind of yarn is used, it should be softly spun to make the sweater comfortable to layer over a shirt and under a jacket.

This sweater starts at the bottom of the back and is knit up over the shoulders and down the front, without a shoulder seam. The edgings for the neck and the armholes are knit as you go, so it is not necessary to go back and pick up stitches and put on edgings after the rest of the sweater is finished.

Materials Needed

Here are the basic materials needed for the sweater:

American double-pointed needles, size #10½, for ribbings
(Canadian size #3, or metric size 6½)
American double-pointed needles, size #11, for rest of sweater
(Canadian size #0, metric size 8)
Handspun yarn about the size of a wooden matchstick
Small size (36) takes about 400 yards
Medium size (38-40) takes about 450 yards
Large size (42-44) takes about 500 yards

(Hint: If you slip the first stitch of every row, it makes a much neater edge and a less conspicuous side seam when sewn.)
Small (size 36 approx.): Using #10½ needles, loosely cast on 56 sts. Knit in ribbing of k 2, p 2, for 3". Change to #11 needles, work in stockinette stitch for 11½" (or desired length to within ¾" of armhole), ending with a knit row. Next row: k 6, p 44, k 6. Next row: k across. Repeat last 2 rows. Then, bind off 2 sts at the beginning of the next two rows, for armhole. Follow standard directions for the rest of the sweater.
Medium (size 38-40): Using #10½ needles, loosely cast on 60 sts. Knit ribbing of k 2, p 2 for 3". Change to #11 needles, work in stockinette stitch for 12½" (or desired length to within ¾" of armhole), ending with a knit row. Next row: k 8, p 44, k 8. Next row: k across. Repeat last two rows. Then, bind off 4 sts

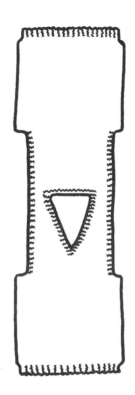

Completed pullover before side seams are sewn.

at the beginning of the next 2 rows, for armhole. Follow standard directions for the rest of the sweater.

Large (size 42-44 approx.): Using #10½ needles, loosely cast on 68 sts. Knit in ribbing of k 2, p 2 for 3″. Change to #11 needles, work in stockinette stitch for 13″ (or desired length to within ¾″ of armhole), ending with a k row. Next row: k 12, p 44, k 12. Next row: k across. Repeat the last 2 rows. Then, bind off 8 sts at the beginning of the next two rows, for armhole. Follow standard directions for the rest of the sweater.

Armhole: (For all of armhole edge, slip the 1st stitch of *each* row.) Knit row: sl 1, k across. Purl row: sl 1, k 3, p 44, k 4. Repeat last two rows until sweater measures 12″ (11″ for size small) from bound-off sts at armhole, ending with a knit row.

Neck opening: sl 1, k 3, p 7, k 30 (this 30 sts can be on #10 sock needles if desired, to give firmer neck), p 7, k 4. Right side: k 11, k 30 (on smaller needle if desired), k 11. Repeat last two rows twice, ending with purl row. Right side: k 15, bind off 22, k 15.

SPINNING FOR SOFTNESS & SPEED   70

Wrong side: k 4, p 7, k 4. Then join another ball of yarn for the other shoulder, and k 4, p 7, k 4. Knitting on both shoulders at once, with two balls of yarn, right side: sl 1, k 11, inc, k 2. Other shoudler: sl 1, k 1, inc, k 12. Purl side: sl 1, k 3, p 8, k 4. Purl side: sl 1, k 3, p 8, k 4. Right side: sl 1, k 12, inc, k 2. Other shoulder, sl 1, k 1, inc, k̃ 13. Continue to increase one st each side of neck opening on knit row (or every other knit row for 2″ if much deeper V-neck is desired) until there are 28 sts on each side of neck opening, keeping 4 sts of garter-stitch at neck edge and at sleeve edge. Work 1″ more, without further inc. End with purl row. Join neckline edges as follows: (on right side) sl 1, k 23, sl last 4 sts of this side onto an extra needle. Lap these last 4 sts of this shoulder over first 4 sts of other neck edge. Knit 1st st of each needle together, then knit 2nd sts together, and 3rd and 4th, forming lap-over of neck facing. Continue across with one ball of yarn. Now, knitting front in one piece, with 1 ball of yarn, keep 4 sts of garter-stitch at sleeve edge, until garment measures same armhole length from top of shoulder to armhole as does the back of the sweater, from armhole to shoulder. Then, cast on (2 sts for small size, 4 sts for medium, 8 sts for large size) at the end of the next two rows. Work (6 sts for small, 12 sts for med and large) edge in garter-stitch for three ridges. Then knit plain stockinette stitch on front until front is same length as back of the sweater down to the ribbing. Change to #10½ needles and work ribbing to correspond to back ribbing. Bind off loosely. Sew side seams. Work in ends of yarn left at neck edge.

To look its best, this sweater should be given a good steaming. Lay it flat, using wet cloth under iron set for Wool temperature; steam with light pressure of iron on wet cloth.

8 Appendix: Tips and Reminders

STORAGE OF FLEECES AND WASHED WOOL

How can fleeces be protected from moths? Moths have a great preference for dirty (grease) wool, especially that which is left in the dark, undisturbed. When wool is sealed up to protect it from moths, there may already be moth eggs and/or larvae in it. In that case, protecting it from moth infestation is being done too late—you are only giving the moths a sheltered place to do their damage. But if the fleeces are from your own sheep, newly sheared, there is no danger. Moths are not known to attack wool when it is on the sheep.

Moths do not go through paper or through tightly woven cotton sacks, although carpet beetles, which also damage wool, can go through cotton. It is not good practice to store wool in plastic, especially grease wool. If the wool is even slightly damp it is not only potentially combustible, but subject to microbiological damage when sealed tightly. Plastic also causes sweating, in the event of temperature changes such as heat during the day and cold at night.

Heavy paper sacks for livestock feed, turned inside out to get rid of grain residue, make good storage for fleeces. If the wool is packed down very tightly, we have found that this discourages moths from damaging anything but the top layer, if there is any damage at all. Fleeces should of course, be packed in well and the bag taped or stapled together at the top. Moderate quantities of fleece are practical for storage in pillowcases, or even in sheets (without holes) sewn into large bags.

The best plan is to keep the raw materials "moving." Store your grease wool less than a year, then wash it. Now it can safely be stored again, this time reduced in weight and bulk. For this second storage it could be sealed into paper or cloth bags until ready to use. Large fiber drums with tight-fitting tops (often discarded by warehouses) are ideal for washed wool or for stored yarn. Seal around the metal lid with masking tape. If the wool is to be used soon, there is no harm in keeping it in large plastic bags, open at the top.

Shaking of fleeces and airing them in the sunlight occasionally can help control moth problems. Twenty-four hours' storage in the freezer is said to kill moth larvae (but who has that much freezer space?).

There are many herbal moth repellent suggestions. Any herbs with a strong smell—such as mint, thyme, or sage—may be something of a deterrent, but they are hardly guaranteed. Moths may avoid them if anything else good to eat is available. Some household insecticides contain an ester of pyrethrolone, which is the active ingredient of flower heads of *Chrysanthemum cinerariifolium*, a reliable insect repellent.

Moth balls and moth flakes are not as harmless or non-toxic as users sometimes suppose. They can be fatal if ingested, so should definitely be kept out of the reach of children and pets. Prolonged inhalation of the fumes can be harmful, as well as handling them without gloves. Moth balls should not be used in plastic bags, as the chemicals in them are not compatible with most plastics and the combination could stain the wool.

The odor of moth balls can be removed from woven garments that have been in storage by hanging the garments over a bathtub containing very hot water with some vinegar in it

PERMANENT MOTHPROOFING

Spinners of tapestry yarns often want to treat their yarn to insure that tapestries made from it will not be damaged by moths.

There are two permanent mothproofers that may be used for this — but both should be used with care.

One mothproofer is registered in the U.S. for industrial/commercial use only, but is hard to obtain in the U.S. It is Mitin® FF High Conc. It is said to be "the least lethal" of the permanent mothproofers. Personally, I have never used a chemical mothproofer, they sound so scary.

The other mothproofer is E Dolan U, Highly Conc., available from Mobay Chemical Company, Dye Dept., P.O. Box 385, Union Metropolitan Park, Union, New Jersey 07083. To be fully informed as to the care and risks in its use, ask for their Safety Sheet #TDS 1086/1.

Some tapestry weavers would rather treat the tapestry after it is woven and ready to go on the wall, using spray mothproofers marketed for household use. This way they do not have to work with yarn that has been treated. Application directly to the tapestry, however, will have to be repeated periodically, particularly on the back.

WOOL WASHING

It is especially important to use washed wool when spinning with one hand. The object in washing is to get out all the gumminess or stickiness so that the wool fibers slip smoothly and easily against each other. The wool will all be washed and rinsed again after it is spun, so this first washing is only to get it clean enough to spin fast and effortlessly.

The ideal arrangement is to spend as little time as possible in washing the largest amount that is convenient. This means less time is spent on drudgery, leaving more time for spinning. If you wash one pound at a time, you are spending nearly as much time on it as it would require to wash twenty-five pounds. You could have done it all at once in little more than twice the time.

Twenty-gallon portable laundry tubs, sold by Sears Roebuck, are a good size to soak and wash about twenty-five pounds of grease wool. For this quantity, fill the tub to within about six inches of the top with very hot water and dissolve in it about ten cups of detergent. This is about equal to a forty-two-ounce box of detergent. Add water softener if your locality has hard water. Pull your fleeces apart and shake a portion at a time, to shake out any seeds or dirt that will come out, then push it down into the sudsy water. Keep putting in the wool until the tub has all that it will hold (up to twenty or thirty pounds) — that is, until you can push down on the top of the wool and not even get a puddle.

Leave the wool soaking in this hot sudsy water for several hours to loosen dirt and grease. Try not to let it cool down. Insulate the tub, or top it with hard foam layer on top.

To remove the wash water, put the wool in net bags and put it through the spin cycle of your washer. Then rinse it in water about the temperature of the slightly cooled wash water. Fill the portable tub half full to rinse half of the wool, then drain out that rinse water (it will be almost as dirty as the wash water) and fill it half full again to rinse the other half of the wool. Just take the wool by double handfuls and souse it up and down in the rinse water, then drop it into a bag and spin out again in the spin cycle. The mesh bag is to protect the washer, not the wool. There is no need to use more than one rinsing, since the wool will be washed again after it is spun into yarn. There is also no need for a wire mesh in the bottom of the tub, as is sometimes suggested, since the tub is too packed with wool for the dirt to sift to the bottom in the washing, and the wool never rests on the bottom during the rinsing.

These twenty-gallon portable tubs have a drain at the bottom to which a hose can be attached; or the water can be emptied into buckets and carried out if you are washing indoors.

The washed and rinsed wool can be dried on wire racks and sorted to color (in the case of dark wools) after it is dry. It is hard to do a color sorting of unwashed fleeces, since you can hardly tell the difference between light gray and pale tan and dirty white until after washing.

WOOL SORTING

It is not necessary to sort a fleece out into different grades, even though there are different grades of wool from the various parts of the same fleece. The difficulty in that kind of sorting is that keeping the parts of the fleece separate requires a lot of extra work; it means that each part must be washed, teased, carded, and spun separately. When you are done, you may still not have enough of any one grade to make whatever you have in mind.

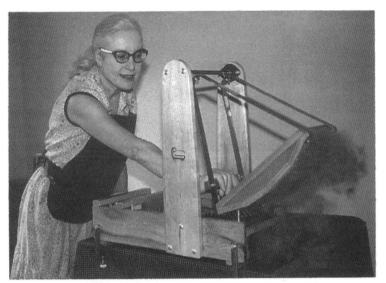

A wool picker makes teasing wool an easy job.

The alternative is to blend the whole fleece together for a large enough quantity of wool to complete a whole knitting or weaving project. This blending is done after washing, during the teasing and carding.

When washing several black sheep fleeces, there is an advantage of doing a *color* sorting after the wool is all washed and dry. You can take the darkest parts of several fleeces for one blended batch of wool, and the brown or gray or tan parts of several fleeces to blend together. This will give you large enough quantities of each color to be useful.

During this color sorting, you may want to set aside any portions, however small, of some unusual shade rather than allowing it to be blended in with a larger amount and thus lost. In weaving or knitting, you can use up a lot of the more plentiful middle gray shades and still have an attractive article if you can brighten it up with some of the more hard-to-come-by colors. Even wisps of color spun into the more dominant shade, although used sparingly, can add a great deal of interest to a garment.

WOOL PICKER USE

This cradle-style picker was detailed in my book *The Handspinner's Guide to Selling*. It does a much faster and more uniform teasing of wool than can be done by hand. In a continuous process, the top card frame of teeth picks up wool from the front slope, and as you swing the cradle, the machine propels the wool from front to rear, where it drops out. A box or basket can be there to catch the teased wool as it emerges. It will process a pound of nice washed wool in about five minutes. Matted wool should be pulled apart somewhat before placing it in the front bin. The 600 needle-sharp teeth do a beautiful job. It must be used with care, and kept out of the reach of children.

Although you may use the machine for either washed wool or grease wool, the gumminess of most grease wool would slow down the process and require more effort to swing the cradle. Also, more vegetation will drop out of washed fleeces during this picking than would come out of greasy wool. Although much of this vegetation will fall through the slits between the bottom card pads, you will still need to brush or vacuum these cards periodically to keep the teased wool cleaner.

Wool does not necessarily need carding afterwards. Try spinning directly from the picked wool for an extremely interesting yarn.

This wool picker is fast becoming standard equipment for the handspinner. It would be most useful to anyone who uses a small drum carder (which always requires wool to be teased for good carding results), and will also protect the card clothing on the drums.

The picker shown here is the Triple Picker made by Patrick Green, 48793 Chilliwack Lake Road, Chilliwack, B.C. V4Z 1A6. It has over 600 hardened tempered steel teeth and sealed-in ball bearings. There is also a positive lock on the swinging cradle, which allows the machine to be carried safely and to be padlocked (padlock included). Be sure to read safety and use instructions.

The height is adjustable so that the picking cradle can be raised for longer wool and lowered for shorter wool. The higher the top cradle, the easier it is to pick. The lower the cradle, the more work it does. The top and the bottom points must not touch. If you feed wool into it too fast, it will not pick as well, and will require more effort. If too much wool goes in at once, don't try to ram it through. Just pull the top cradle towards you, remove a handful of wool (carefully) and refeed it a little more slowly.

There are clamps to attach the picker to a table, to keep it securely in place as you work.

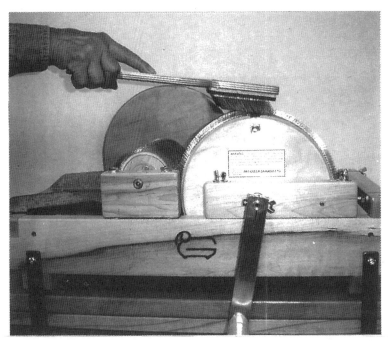

The burnishing tool in use. Long flexible wires on a comfortable wooden handle. From Pat Green Carders.

THE BURNISHING TOOL

This innovative hand-held tool keeps your carder's teeth sharp and bright. At the same time it gives you a better and thicker batt by providing greater working power than could otherwise be attained with a two drum carder.

By lightly burnishing the batt, intermittently as you card, it actually amplifies the working power of the carder drums.

While originally developed to assist in the processing of exotic fibers such as alpaca, bunny, cashmere and kid mohair, it has proven useful for all wool fibers, producing a smoother and thicker batt which is easier to peel off. Fly-away fibers and static fluff are also better controlled by use of the Burnishing tool. It is one of the accessories included with the Supercard, but is equally valuable for use with any carder.

DRUM CARDERS

The number of carding teeth per inch is not the prime indication of the capability of a carder. More important to its versatility is the fineness and flexibility of the tooth wire, the location of the "knee" (the bend in the wire), the tooth shape and angle in relation to the foundation, and the firmness of the cushioned backing. Very few hand-turned carders are perfect for all fibers, no matter what the tooth wire design.

The carding ratio is the most important factor to consider in a carder. This is the relative difference in speed of the two drums, and the ideal ratio is not the same for the major fiber classes. For instance, too high a ratio is a disaster with the fine crimpy fibers like Merino and Rambouillet. They tend to stretch out, snap back and make nasty noils.

The carding drums should be adjusted as close as possible to each other, without touching. Noise of teeth scraping against each other indicates they are damaging each other, as well as being rough on the fiber.

As to the type of drive mechanism, a belt drive seems safer than sprocket and chain. Chain drive will not slip if you run into matted locks, thus the chance of bending carder teeth. It also makes it difficult to secure the best drum ratio for good carding and blending. There is a simple crossed belt arrangement which is an economic design, but this suffers continual friction on the belts as they turn.

The main consideration in a carder is: does it work well with your fiber choices? Before purchasing, if possible, try out carders belonging to several friends — with your own fibers.

MOTORIZED CARDERS

For serious spinners, and for those contemplating a small and pleasurable home business, a motorized carder is a necessity.

Why motorized? You get better carding with a motor since it goes at a steady speed, while hand turning means stopping, starting, slowing down, speeding up, all having an effect on the batt. Power also allows the use of both hands for feeding fibers and arranging color blends. A motor also permits a variable carding ratio arrangement, making the carder completely versatile. While a two speed or three speed option is helpful, an infinite dialed ratio, as featured on the Supercard, is ideal. This is one of the most important advances in tabletop carders in the last fifty years.

The three-drum Supercard is the ultimate in versatility

The Supercard. Carding excellence in batts or roving! Made by Pat Green Carders Ltd.

and gives exceptionally high quality results with ALL fibers. Its much larger main cylinder gives a greater arc of contact between drums, which is where the carding actually takes place. The little "worker" drum above the infeed roll is important for added working power and conditioning of the batt as you card. Since the main cylinder is free-wheeling when the motor is turned off, this makes it easy to use its Roveguide system to take a roving direct from the drum, if you prefer roving to its three-foot-long batt.

As an added convenience, a powered carder does not require table clamps, it just sits there nicely and purrs at you.

The next size up is the four-drum Elsacard, a logical progression as a little fiber business grows. Still tabletop, but four times the size and four times the production of the Supercard, capable of batts (5 foot long!) or roving. Great for feltmakers desiring large batt size, or for production spinners enjoying a longer roving and increased output.

WOOL "CONDITIONING"

The most important factor in improving the quality of spun fibers and adding to the ease of spinning, is their preparation. Fiber selection, washing, conditioning, picking and carding, all play their part.

The use of water-soluble oil can make difficult wool easier to use. Diluted and sprayed lightly on the wool the night before carding, can make some wools easier to card without damage. Oil can be added to rinse water for fleeces that have a tendency toward matting.

My personal preference is a conditioning mist of warm water containing either some laundry anti-stat liquid, or some cream hair rinse. This type of conditioning avoids oily odor, and the need for soap-and-water washing to remove it. Also, it cuts down greatly on the static problems that occur with some fibers, especially in cold, dry weather conditions. If you are in a hurry, mist the wool, put it in a black plastic bag in the sunshine, for faster results.

This conditioning can be done either before or after use of the picker. If static is a problem, then conditioning after picking is probably the most helpful sequence.

JOINING ONTO THE LEADER

For the beginner, having a knot in the end of the leader will help the wool to cling to it when joining on to start spinning. To make sure the joining will not pull off, spin out a good length of well-twisted yarn, while still holding onto the leader with the hand closer to the orifice, before allowing the yarn to wind onto the bobbin. It is then not so apt to pull loose when tension is put on it in starting to spin the next section of yarn. The joining is by then well wrapped around the bobbin, due to the length of yarn that was initially spun to the leader.

If you're spinning from mill-carded sliver or combed top, which can be quite slippery, the end of the leader can have a knot into which the beginning fiber supply can be caught.

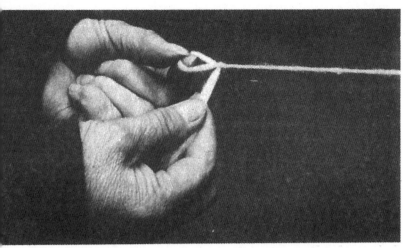

Wool clings well to a knot in the leader. It can also be caught through a knot such as this one.

JOINING ON

It is possible to have had a year's practice in spinning but little in joining. When you stop spinning to reach for more fibers, join them on, and continue spinning, you are only getting practice at stopping and starting. The real practice in joining happens when you join "on the run" without stopping.

There is one way to get the equivalent of a year's practice in joining, all in one skein, just by doing a whole skein of joining. Put a lot of little puffs of wool on a table or a bench beside you, within easy reach. It helps if they are puffs of several different colors, to make it more interesting for you. Then spin (as fast as you can), grabbing at those little bits of wool and joining them on while you are spinning. Depend on the dexterity of your hands to decide how they will handle the joining, by forcing them to work fast. Don't worry how the yarn looks. You will find, when it is done, that it is interesting enough so that, if you were selling yarn, that would be the skein the customer would choose.

In the joining on of each new fiber supply, one standard bit of misinformation says, "Taper your yarn off to a fine point and then lap the next rolag over this" to join on. If you intend to ply your yarn, this kind of joining might possibly be adequate. However, it is unsatisfactory in single ply and would probably pull apart when you try to wind off the yarn into a skein, or wind

Feathering method of joining, done with one hand.

the washed skein onto the blocker, or make a warp chain of it for weaving.

On the other hand, if you do not spin out quite to the end of your fiber supply, instead leaving a few inches attenuated but completely untwisted, then it is a simple matter to fan out this untwisted mass, lay the ends of your next fibers against it, and continue treadling. The twist will cause the fibers to spiral together, in the same manner as if there were no interruption and joining. A good joining is of unspun fibers to unspun fibers. If your yarn breaks and you do not have a fluffy unspun wisp to join against, then feather out the end of the yarn that you want to attach to, so that it is again unspun and will join securely. It is a more perfect join, also, if the fibers are *drafted out* together, not just *twisted* together.

The above manner of joining, when done at fast speed, does not bear much resemblance to the same action when done in slow motion. The only explanation I can give is that the speed of the spiraling fibers catches in the fibers from the newly added supply in a different (and better) way when all are moving faster.

There are actually two principles of joining. The one just described is a drafting out of the previous fiber supply together with the new fiber supply. When done slowly, it could be called

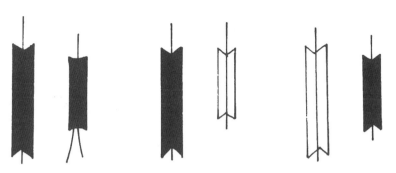

Pulley power arrangements, left to right: double belt, bobbin lead; single belt, flyer lead; single belt, bobbin lead.

fanning out. The other way is the feathering together of the new supply into the last of the previous supply, usually done with two hands when working on a flyer spinning wheel and with one hand when working on the great wheel. This is like shuffling cards, catching fibers from the previous wool supply along with fibers from the new supply, until they are joined.

SPINNING WHEEL PULLEY ACTION

The three kinds of pulley power arrangements are:
1. Double belt, bobbin lead (pronounced lēd). Has power on both the flyer pulley and the bobbin pulley. The flyer causes twist and the bobbin causes pull (with a slipping-clutch action when it is not pulling in).
2. Single belt, flyer lead (Scotch tension). Has power on the flyer pulley and the brake on the bobbin pulley. The flyer causes both twist and pull (slipping brake).
3. Single belt, bobbin lead (Indian-head tension). Has power on the bobbin whorl and the brake on the flyer shaft or pulley. The flyer causes twist and the bobbin causes pull (slipping brake).

SPINNING STYLES

The Inchworm

"Inchworm" is the common term for the spinning method most used by beginners. With this method the wool is less apt to pull apart than with other ways of spinning.

The hand further from the orifice holds the wool supply,

while the hand closer to the orifice inches out the fibers into yarn. Impeccable yarn can be spun this way, but these particular hand movements cannot be speeded up appreciably. The inchworm is a useful technique for beginners because it enables them to gain confidence in controlling the wool. Then they can progress directly to a one-handed style by allowing the spinning wheel to take over the function of the hand nearer the orifice. The changeover can start with short lengths of yarn drawn out and allowed to pull onto the bobbin, then advance to a longer sweep and more fluid movement.

The Short Draw

For this technique, the hand positions are the same as for the inchworm, except that in this case the hand closer to the orifice does not move. With a pinching action it controls the twist, while the hand further from the orifice moves, drafting the yarn in short lengths and allowing it to wind on. In using antique wheels with a shallow pulley ratio, this technique works very well. It is most often done as an intermittent drafting and winding on, but can be refined into a continuous movement that is as fast as the long draw.

The Long Draw

Here, the hand closer to the orifice controls the twist, allowing it to build up, and letting it into the drafting fiber supply as the hand further from the orifice attenuates the slightly twisted wool, pulling against the firmness of the hand that controls the twist. The twist is allowed into the drafting fibers a little at a time. As the desired size of yarn is attained, the balance of the built-up twist must be released and the yarn allowed to pull quickly onto the bobbin. The hand nearer the orifice again pinches off the twist, then releases it as needed, while the hand holding the fiber supply again drafts out the yarn, usually to the length of the outstretched arm. This is a difficult technique for beginners, inasmuch as it is hard to judge the amount of twist needed and keep that amount constant throughout the skein.

Although this consists of alternate drafting and winding on, it can be a very fast way of spinning when done with a spinning wheel that really snaps in the yarn.

The Unsupported Long Draw

This is also called "double drafting." It is a good technique for heavy yarn, although there is no reason why it should not be done with finer yarns too. It consists of the attenuation of an almost untwisted (to begin with) fiber mass, as the twist is entering into it. The hand nearer the orifice helps with the drafting after a slight amount of twist has entered the unsupported fibers. Be sure to work with hands far apart. The hand holding the fibers will be doing the actual drafting, although the feeling will be that of stretching out the fibers between both hands while the hand nearer the orifice is pinching, letting the twist through, and at the same time being a firmness to be pulled against. The yarn will look quite thick-and-thin at first, but as you continue drafting, the thicker (and less twisted) spots will quickly even out into about the same size and twist as the thinner places. Ideally, too much twist must not be inserted before the wool has been drafted to the desired size and texture. If too much twist seems to be entering the yarn, this must be controlled by pinching off the twist with the hand nearer the orifice. If then necessary, more unspun fiber supply can be released, so the overtwist can be spread out over more yarn area and thus equalized.

This releasing and drafting of an unspun fiber mass can be performed with a rhythm that just invites speeding up, and is probably the fastest method for spinning a medium-heavy yarn. What is required for speed is wool that is washed clean and nicely carded, a fast spinning wheel, and a very rhythmic drafting. If these requirements are met, the result will be a very evenly spun medium-heavy yarn.

YARN BLOCKING

When skeins of yarn are hung with a weight, you can look at the yarn and see that part of the skein is actually weighted and part of it is not even under tension. It is impossible to hang it in such a manner that all of the strands are equally weighted. Since they are not, you get a skein in which some parts are blocked while other parts remain unblocked. Because of its random areas of elasticity, the yarn when used gives unreliable results such as

Louët blocker with skeins drying on it. Pegs at both ends of the blocker reverse and point outward (note the ones at left, turned inward) so the blocker can also function as a warping reel.

irregular selvages in weaving and the appearance of uneven tension in knitting.

The alternative is blocking, which assures that all parts of the skein are dried under an even tension. Each skein is wound onto the blocker under a light tension, and left on only until dry.

This blocking process is of most benefit to singles, but in some instances can be useful for plied yarn. If you have a novelty yarn and wonder if blocking would improve it, give it a try. One yarn that should not be blocked is thick-and-thin slub yarn, where you want the slub to be soft and fluffy. In that case, blocking would tend to flatten the slub.

Does this block out some of the elasticity? Yes, it blocks out the *excess* elasticity, which is a problem no matter what you do with the yarn, making it difficult to figure a stitch gauge or to knit a sweater that will not stretch out of shape when you wash it. Excess elasticity can give trouble in weaving, making it hard to estimate the amount to allow for take-up and shrinkage in both warp and weft.

To block a skein of yarn, just wash and rinse it as usual, but instead of hanging the skein to dry, put it on a swift and tie one end to the blocker. Wind the whole skein onto the blocker, then tie both ends around the skein before you put the next skein on the blocker. This is so that the skeins are set apart and do not overlap, and are tied off ready to be removed the next day. When the skeins are dry, lift the blocker out of its cradle and slip the skeins off the ends. In fact, the blocker frame does not have to stay in the cradle while the skeins are drying; it can be lifted out and taken to a warm room, or outdoors in a breeze, to dry.

The Louët Sales Spinning Wheel Company offers a combination blocker and warping reel. Operated as a blocker, it allows a handspinner to dry a dozen fifty-four-inch skeins at once. Used as a horizontal short-warp reel with the pegs for the cross turned outward, it can hold a warp of up to twelve yards long. (The length is of course affected by loom width, chain size, and yarn thickness, so it could be longer or shorter than this.)

For using the device as a blocker, special instructions are included which allow for relaxing the tension on the yarn. The maker says that this slight relaxation of the yarn, before drying, noticeably improves its feel.

To do this blocking, put your first skein on the swift or skein-winder. (Blocker pegs should face inward, so as not to obstruct skeins.) Tie off the end of the first skein on one of the blocker's horizontal yarn bars, about three inches in from one end of the blocker. Now wind the skein on and make a skein on the blocker. When done, wrap the end of the yarn around that skein and tie off. Do the same with the end that was tied to the blocker.

Put a fresh skein on the skeinwinder or swift, and repeat, placing each new skein towards the middle of the blocker and not overlapping the skeins. When you arrive at the "crown" halfway along the yarn bars, start three inches in from the other end, and wind skeins on, towards the middle again.

To relax tension, slide the first skein towards the end of the drum about two inches, and move all other skeins toward the end by the same amount. (If you don't move them, shrinkage will increase the tension.) What this does is relax the yarn slightly. You can now leave the skeins to dry. When they are dry, lift the drum out of the base and stand it on one end. Slide half the skeins off over the upper end of the drum, turn the drum over and remove the others.

Handle the blocker carefully when it is up on end. Dropping it or leaning on it heavily could bend the axle bolts. Simple blocker plans available from Pat Green Carders. Send $1.00 and a long stamped envelope.

WARP SIZING

Sizing a handspun warp will give it extra strength; it will also reduce wear from the heddles and reed during warping and during weaving. It prevents the fuzzing of the yarn that can make a "sticky" warp (one that clings to itself) and keeps the selvages from fraying. If there are any weak joinings, sizing lessens the chance of their pulling apart during weaving.

Sizing is done after the warp yarn is washed and blocked and wound into a chain, and before the chain is warped onto the loom. The sizing solution is a foamless natural hide glue. The dry granules are soaked in cold water until they absorb the water, then boiling water is added to liquefy the soaked granules. The formula, which can be multipled for a larger quantity, is three tablespoons of granules in one cup of cold water, with one cup of boiling water added after the soaking.

The dry warp chains are immersed in this hot solution, then wrung out through a clothes wringer. Old-fashioned wringers are often found in second-hand shops (not antique shops, which are too expensive). New ones are available, but they are also expensive. An old washing machine with an electric wringer is good, but requires two people — one to keep the warp from wrapping around the wringer rollers as it comes through. A roller mop-wringer could be used if you can't locate a clothes wringer, but the warp would have to be contained in a mesh bag to avoid damaging it by pulling it forcibly through the mop wringer. A plastic salad-spinner gets out enough of the excess sizing from each chain, if you spin it long enough.

After the sized warp chains are wrung out, the warp is unchained and hung horizontally to dry completely before tying onto the loom.

Warp sizing granules available in half-pound packages from Susan's Fiber Shop, N 250 Highway A, Columbus, WI 53925.

DUMMY OR PERMANENT WARP

This is a short warp made of cotton carpet warp, threaded through the heddles and reed. It should be the right number of threads per inch, and of the same total number of threads, that you plan for your project. Tie your sized handspun warp chains onto these carpet warp ends, one by one, using a weaver's knot. When all are tied on, wind it onto the loom, through the reed and heddles. As you wind it on, allow enough remaining handspun warp in front of the reed to be able to tie onto your front apron. By using the dummy warp you avoid the waste of handspun that would occur with other ways of warping, and eliminate the need to rethread the heddles and the reed for each subsequent warp of the same threading.

VEGETABLE DYE SHORTCUTS

1. Dye in larger quantities, in order to reduce the time spent per pound of wool being dyed.
2. Dye fleeces rather than yarn. With fleeces, the dyeing does not have to be done with precision to avoid streaking and uneven color. Wool fleeces need not even be stirred; uneven dyeing gives much more life to the color after blending or teasing with a picker, or carding.
3. Use mono-mordanting, in which the mordant is in the same pot with the dye. With this method, wool can be washed and rinsed and then go directly into the dye bath without separate mordanting and drying and re-wetting.

There are several advantages in having the mordant in the dye bath. It cuts down on the time element in dyeing, as well as some of the cost. It also eliminates double boiling of the wool, thus safeguarding quality and resulting in brighter colors.

Actually, in my own mono-mordant dyeing, I used only one mordant combination. That was alum (potash alum) and cream of tartar (bitartrate of potash), which I found to be most ecologically safe in a farm situation and also nontoxic in handling. The Center for Occupational Health in Hamilton, Ontario, says that alum is relatively nontoxic unless consumed in large amounts, and cream of tartar has low toxicity and is laxative if consumed in large quantities. Other commonly used mordants are much more dangerous: the lethal dose for an adult is one teaspoon of copper sulphate, seven drops of potassium dichromate (chrome) or one ounce of ferrous sulphate, and these amounts would all be considerably less for children.

SOURCES

CARDERS, PICKERS & BLOCKER PLANS
Pat Green Carders Ltd.
48793 Chwk. Lake Road
Chilliwack, B.C. Canada V4Z 1A6
604-858-6020

A FEW SPINNING WHEEL SOURCES
Schacht Spindle Co. Inc.
5101 Ben Place
Boulder, Co. 80301
303-442-3212

Louët Sales
R.R. 4
Prescott, Ontario, Canada K0E 1T0

Lendrum
403 Millhaven Road
Odessa, Ontario, Canada K0H 2H0

WOOLEE WINDER
Robert Lee and Son
Box 941
Oregon City, OR 97045
503-810-1388

WARP SIZING
Susan's Fiber Shop
N 250 Hwy. A
Columbus, WI 53925
920-623-4237

MOTHPROOFERS
Hillcreek Fiber Studio
7001 Hillcreek Road
Columbia, MO 65203
573-874-2233